PROGRAMMING THE STATISTICAL LIBRARY

PROGRAMMING THE STATISTICAL LIBRARY

by James H. Hogge
Associate Professor of Psychology
George Peabody College for Teachers

First Edition

AUERBACH®
publishers

philade|phia
new york
london

AUERBACH Publishers Inc.,
Philadelphia 1972

Library of Congress Catalog Card Number: 72-79974
International Standard Book Number: 0-87769-136-3

First Printing

Printed in the United States of America

Contents

Preface   vii

1. Introduction   1
The Need for Library Programs; Conversion versus Creation; General Characteristics of a Library Program

2. Control Cards   5
Generalizing a Program; Alternative Control Card Schemes; Subject Identification Fields

3. Documentation   35
Assumptions about Users; Functions of Documentation; Examples of Documentation; Summary

4. Programming the Small Computer   59
Problem Segmentation; Storing Matrices Efficiently; Summary

5. The IBM 1130 Disk   81
Use of Disk for Temporary Files; IBM 1130 Disk Monitor and Library Programs; Use of Disk for Permanent Files; Summary

6. Use of IBM 1130 Commercial Subroutine Package   107
CSP Routines for Input/Output; Packing Alphanumeric Information; Summary

7. Library Program Standards   123
Input Standardization; Output Standardization; Documentation Standardization; Source Deck Standardization; Summary

8. Case Studies of Illustrative Library Programs   135
Pearson Product-Moment Correlation; Frequency Tabulation for Single-Column Variables; Summary

Appendix A: *Listings of Subprograms*   167
Appendix B: *Peabody Computer Center User's Manual*   185
Index   203

Preface

It is the intent of the author to provide sufficient instruction in a short volume to permit a modestly skilled FORTRAN programmer to create highly polished library programs for statistical analysis. Accordingly, this book is intended for use by individuals in education, the behavioral sciences, or any other area in which computationally naive researchers routinely perform statistical analyses.

There are several important things this book is *not* intended to do.

First, it is not intended to provide an introduction to the syntax of FORTRAN or operating systems. The reader is assumed to be reasonably facile with respect to both the language FORTRAN and the operating system of the computer for which he plans to create library programs.

Second, no attempt is made to introduce the reader to statistics. Instead, it is assumed that the programmer attempting to create statistical library routines will draw either upon his own training in statistics or on the intimate involvement of a consultant competent in statistics and sensitive to the needs of the users for whom the library is intended.

Third, the reader of this book will find almost no algorithms for the implementation of statistical techniques. This book is devoted instead to the presentation of programming concepts that differentiate between a statistical *library* program and a statistical *special purpose* program. The reader requiring algorithmic assistance is referred to the works of such authors as Borko, *Computer Applications in the Behavioral Sciences* (Prentice-Hall, 1962); Cooley and

Lohnes, *Multivariate Data Analysis* (Wiley, 1971); Horst, *Factor Analysis of Data Matrices* (Holt, Rinehart and Winston, 1965); Pennington, *Introductory Computer Methods and Numerical Analysis* (Macmillan, 1965); Ralston and Wilf, *Mathematical Methods for Digital Computers*, Vol. I (Wiley, 1960) and Vol. II (Wiley, 1967); Sterling and Pollack, *Introduction to Statistical Data Processing* (Prentice-Hall, 1968); and Veldman, *FORTRAN Programming for the Behavioral Sciences* (Holt, Rinehart and Winston, 1967).

Fourth, although this volume is intended to present techniques and subprograms useful in the production of library programs, no attempt is made to present a complete library of programs covering the statistical waterfront. Readers desiring such a collection may be particularly interested in the books by Cooley and Lohnes, Horst, and Veldman.

Because an increasing number of such small-scale computing systems as the IBM 1130 are found in computational facilities intended for general use, special emphasis has been placed upon efficient utilization of systems relatively limited in core memory. Nevertheless, the concepts presented are applicable to large-scale computing systems as well. In order to minimize conversion problems, special care has been taken to present FORTRAN programs utilizing only basic features of the language.

The author would like to express his appreciation to a few of the individuals who provided assistance during the development of this book. Colleagues who read the manuscript and made helpful suggestions were Richard Gorsuch, Martin Judd, Howard Sandler, and Jim Shackleford. In addition to reading the manuscript and making suggestions, Jamal Abedi, Joe Gaida, Joe Harm, Ed Roby, and John Spollen attempted to put the control card schemes and subprograms of the book to use in the development of actual library programs.

Special thanks are in order for Nell Ayers, who typed the manuscript, and Pat Marsh, who helped assemble it. Their efforts kept the project running smoothly.

J. H. H.

February 25, 1972

PROGRAMMING THE STATISTICAL LIBRARY

1 Introduction

The creation of FORTRAN marked a major advance in the movement from problem expression in machine terms to problem expression in human terms. Program development in the early days of computing required knowledge of highly machine-oriented languages almost totally devoid of similarity to the language of any applications area. Before FORTRAN, man approached the machine on its terms, but FORTRAN now permits communication with the computer in terms closely resembling those in which "scientific" problems are expressed. Not only has it shortened the time required for computer language acquisition but it has dramatically reduced the time consumed in program coding. Why, then, are program libraries needed?

THE NEED FOR LIBRARY PROGRAMS

Education and the behavioral sciences do not abound with "quantitative types." Most statistical analyses are performed by researchers who regard statistics as only a tool. Concerned primarily with substantive research questions, these individuals desire only such forays into methodology as may be required for respectable rigor. This is not to say that the typical researcher knowingly casts his experiment in an inadequate design or slights his statistical analyses. Quite to the contrary, most research projects are piloted by conscientious investigators who want to collect interpretable data amenable to powerful analytic techniques. Nevertheless, it is essential for the data processing consultant to recognize the patterns of interests, skills, and aversions that characterize his clients.

The typical educator or behavioral scientist probably has not mastered FORTRAN. It is possible that he is minimally prepared with respect to training in statistical analysis, and that he becomes uncomfortable at the thought of approaching the computer. By the time the beleaguered researcher overcomes his fear and presents himself at the office of the data processing consultant, he is a client in crisis. This unfortunate state of affairs has led more than one computer center employee to comment upon the disagreeable humor displayed by statistical program library users. If, however, the stressfulness of the user's situation is recognized, his behavior is understandable. The data processing consultant's role becomes one of reassuring the researcher and getting him to the computer with as little discomfort as possible.

Even the comparatively rare FORTRAN programmer needs a statistical program library. The variety and complexity of statistical analyses in common use in education and the behavioral sciences preclude the development of new computer routines for each new set of analyses. Instead, the availability of software "on the shelf" can permit a significant savings in time and effort for even the sophisticated investigator. From the standpoint of allocation of research project resources, it makes little sense for personnel to spend vast amounts of time reinventing the wheel.

CONVERSION VERSUS CREATION

When the need for a particular analysis arises, the typical statistical program library user will peruse existing resources to ascertain whether an appropriate program is available at his institution's computer installation. If no applicable routine is available, the researcher is more likely to alter his research design than to risk being required to confront data analysis using a desk calculator. It is this behavior that imposes an obligation upon computer centers to make available a comprehensive statistical program library. Once habituated to statistical analysis using the computer, the researcher is not easily weaned. Just as the availability of instruments for the measurement of constructs determines, to a considerable extent, the substantive directions of research, access to computational support dictates the popularity of various statistical techniques.

A general recognition of the need for statistical program libraries has resulted in the creation of sets of programs at computer facilities throughout the country. The computer center of almost any major college or university has programs available for "bread and butter" statistical analyses. In addition, most major vendors of computer hardware feature statistical software among their wares. Accordingly, recognition of the need for a particular statistical routine should be followed by a search for an existing program. Such journals as *Behavioral Science* and *Educational and Psychological Measurement* include sections devoted to dissemination of information concerning existing software. Most of the programs described have been written in FORTRAN. The ubiquity and standardization of the FORTRAN language often make it possible to transfer a program from one computer to another with minimal conversion difficulties. The feasibility of conversion of an existing program should be carefully considered before an effort to create a new routine is launched.

Statistical analysis is not an especially stable area of endeavor, however, and new techniques do appear from time to time. Also, an investigator's individuality may manifest itself in the requirement of particular program features not found in existing programs. Finally, a need for the creation of a new program may be occasioned by major differences in hardware capabilities. If, for example, one system has much more core storage than another machine, it is unlikely that a program written for the first computer can be easily converted for use on the second.

Although the choice between conversion and creation is often clear-cut, there are numerous occasions on which it is not. In fact, the process of conversion of a program could entail such extensive changes that creation of a new routine might involve less effort. It is often considerably more difficult to immerse oneself in a program of someone else's creation than it is to generate original FORTRAN code. Only experience can assist the programmer in deciding how to add a needed program to his library. Hopefully, this book will help him acquire that experience.

GENERAL CHARACTERISTICS OF LIBRARY PROGRAMS

A library program is a routine written in such a way that it is flexible enough to accommodate a relatively large class of problems. In general, an increase in the flexibility of a program (and the variety of problems it can handle) is accompanied by increased complexity from the user's point of view. The sophistication of the user and the class of problems determine the optimal balance between flexibility and simplicity.

The specification of information required to describe a particular problem for processing by the library program is accomplished using *control cards*. For example, a program to compute Pearson product-moment correlation coefficients among a set of variables measured for a group of subjects might be designed so that the user need only specify the number of variables, the number of subjects, and the arrangement of the data in each subject's card file. The number of variables and the number of subjects are often referred to as the problem *parameters*, and the arrangement of the data in each subject's card file is called the data *format*. It is quite important to standardize control cards as much as possible, thus making it easier for the user to move among the programs in the library. The adoption of consistent jargon can further expedite the user's understanding of library programs he has not used before.

The usual naivete of the statistical program library user and the paucity of consultative staff at most computer centers occasion a need for program *diagnostics*—error messages identifying the sources of difficulty in the event of user error. It is possible for the programmer to anticipate most common user errors and to include tests in the program to detect errors before they precipitate electronic disaster. In no event should the user be left with the impression that he has committed so heinous an error that he has reduced the computer to mute disbelief. Error messages themselves should clearly pinpoint the source of trouble. For example, the computer response

INVALID PARAMETER SPECIFICATION

tells the user considerably less than

INVALID NUMBER OF VARIABLES

A library program is of minimal usefulness if only the priests of the computer center know how to use it. If documentation is too sparse or unclear, intercessory activities on the behalf of users will demand the maintenance of a sizable priesthood. On the other hand, requirements for consultative staff are minimized when care has been taken to produce thorough documentation. Programmers are notorious for their predilection to lose interest in a program after they have written it. The production of documentation is anticlimactic to the programmer who has just emerged victorious from yet another encounter with the digital dragon. His post-conquest inattention can be avoided if he is required to produce the write-up for a program *before* he writes the program itself. This practice can also serve somewhat the same function as flowcharting in program planning.

Although library programs have been written in languages other than FORTRAN (for example, ALGOL or BASIC), FORTRAN continues to dominate the scene in statistical analyses. Just as objectivity is sought in an attempt to make scientific inquiry public and exportable, so should the choice of computer language reflect a desire to make a program useable by as many members of the scientific community as possible. As mentioned earlier, the ubiquity and standardization of FORTRAN recommend it as the present choice for implementation of statistical procedures.

2 Control Cards

As was mentioned in Chapter 1, control cards are used for communication between the user and the library program. In this chapter we shall examine several alternative control card schemes and study their implementation in FORTRAN.

GENERALIZING A PROGRAM

Suppose we have ten test scores for each of 25 subjects, and we have punched our data according to the following layout:

Column	Entry
1-5	Identification
6-7, 8-9, etc.	Ten two-digit test scores

Using these data as test data, we shall trace the development of a library program for the computation of means. Successive versions of the program will incorporate various refinements and library program concepts.

Version I

Figure 2-1 contains a listing of a simple program to compute and print the means of the ten tests for our specific data set. Although our program would

```
C
C        ROUTINE TO COMPUTE MEANS - VERSION I
C
         CIMENSION SUMX(10), X(10)
C
C        CEFINE I/O DEVICE NUMBERS.
C
         KARDS = 2
         LINES = 5
C
C        INITIALIZE SUMX.
C
         CO 5 I = 1,10
      5 SUMX(I) = 0.0
C
C        INPUT DATA FOR 25 SUBJECTS AND ACCUMULATE IN SUMX.
C
         CC 15 I = 1,25
         READ (KARDS,10) (X(J), J = 1,10)
     10 FORMAT (5X, 10F2.0)
         CO 15 J = 1,10
     15 SUMX(J) = SUMX(J) + X(J)
C
C        CCMPUTE MEANS.
C
         CO 20 J = 1,10
     20 X(J) = SLMX(J) / 25.0
C
C        PRINT RESULTS AND EXIT.
C
         WRITE (LINES,25) (I, I = 1,10), (X(J), J = 1,10)
     25 FCRMAT (6HOMEANS // 10I11 / 3X, 10F11.3)
         CALL EXIT
         END
```

Figure 2-1. Routine to Compute Means, Version I

process only 25 data cards punched according to the layout described above, it does have one feature that expedites conversion for compilation and execution on differing computer systems. The use of integer variables for specification of input/output device numbers is permitted by FORTRAN IV. In our program we have used the variables KARDS and LINES to represent the device numbers for the card reader and line printer, respectively. Rather than supply actual device numbers in subsequent READ and WRITE statements, we insert the variable names KARDS and LINES. If we later need to change device numbers in the process of converting our program to a different machine with different device number assignments, a change of only two cards is necessary. The benefits of this ploy are not especially apparent in a simple routine like the one above, but the payoff in a lengthy program containing many input/output statements is considerable.

Another important concept was utilized in the design of the program. It is tempting to conceptualize phases of processing in a statistical routine as consisting of input of data, calculation of results, and printing of results. It would seem consistent with this scheme to read the data into computer memory before proceeding with calculations, but by combining some of the calculations with data input, we can vastly reduce the memory requirements of our program. Many statistical procedures (e.g., analysis of variance and correlation) require the accumulation of sums (ΣX, ΣX^2, ΣXY, etc.), and these sums can be accumulated during data input. Accordingly, only a ten-element array SUMX is needed for the sums in our program, and a ten-element array X is used to hold each subject's data long enough to add his scores to the appropriate sums in SUMX. If we had read all data into memory at once, a two-dimensional array with a total of 250 elements would have been required. This is an insignificantly small memory allocation on most computers, but not if we consider a case with 100 scores for each of 10,000 subjects. An array to hold all data in memory at one time would consist of a million elements, a requirement exceeding the core storage available on all but the very largest of computers. On the other hand, exploitation of the technique of "running summation" used in our simple program would require only two 100-element arrays. The re-definition of the array X in the DO 20 loop eliminates the need for a separate array to contain the computed means.

Finally, it is worth noting the use of the running subscript I by itself in the WRITE statement to generate integer headings for the output. These headings would be thought of as variable numbers corresponding to serial order of input of the scores.

Version II

This version of our routine to compute means appears in Figure 2-2 and incorporates some generalizations of Version I. The integer variables NV and NS have been added to the program so that differing sets of data may be processed. The number of variables, NV, may be defined as any value between 1 and 100. Since the dimensions of SUMX and X must correspond to the maximum number of variables, they have been increased to 100. The number of subjects, NS, is effectively unlimited. The use of NV and NS throughout the program instead of constants, as in Version I, permits the user to adapt the program to varying sets of data by changing only the values assigned to NV and NS and the specifications in FORMAT number 5.

The output of the program has been changed so that the array X (containing the means) is printed in ten-element segments. As in Version I, each mean is labeled with an integer corresponding to serial order of input.

Although Version II is computationally adequate as a program to compute the means of several variables, it lacks the attributes of a good library program. The user of Version II has to handle a sizable deck of cards in which order is critical. Unless the user is acquainted with FORTRAN, he is likely to be intimidated by the requirement that he change program statements; furthermore, he will be utterly helpless if he acquires a copy of the deck with cards out of order or missing. In short, there are many things to go wrong, and no provision for nontechnical feedback to identify the source of trouble. Our task, then, is to improve the interface between the user and his computational support.

```
C
C      RCUTINE TC CCMPUTE MEANS - VERSION II
C
       CIMENSION SUMX(100), X(100)
C
C      CEFINE I/C CEVICE NUMBERS.
C
       KARCS = 2
       LINES = 5
C
C  *** CHANGE THE NEXT THREE STATEMENTS TO FIT YOUR DATA.
C
       NV = 10
       NS = 25
     5 FCRMAT (5X, 10F2.0)
C
C      INITIALIZE SUMX.
C
       CC 10 I = 1,NV
    10 SUMX(I) = 0.0
C
C      INPLT DATA FOR NS SUBJECTS AND ACCUMLLATE IN SUMX.
C
       CC 15 I = 1,NS
       READ (KARCS,5) (X(J), J = 1,NV)
       CC 15 J = 1,NV
    15 SUMX(J) = SUMX(J) + X(J)
C
C      CCMPUTE MEANS.
C
       SN = NS
       CC 20 J = 1,NV
    20 X(J) = SUMX(J) / SN
C
C      PRINT RESULTS ANC EXIT.
C
       WRITE (LINES,25)
    25 FCRMAT (6HOMEANS)
       CC 50 I = 1,NV,10
       IF (I + 9 - NV) 35,35,30
    30 II = NV
       GC TO 40
    35 II = I + 9
    40 WRITE (LINES,45) (J, J = I,II)
    45 FORMAT (/ 10I11)
    50 WRITE (LINES,55) (X(J), J = I,II)
    55 FORMAT (3X, 10F11.3)
       CALL EXIT
       ENC
```

Figure 2-2. Routine to Compute Means, Version II

One way to reduce the chances of user error is to eliminate as many potential errors as possible. By minimizing the number of cards in the deck the user handles and the extent to which he is required to modify the program itself, we can head off considerable difficulty. Control cards permit the user to supply information required for processing without modifying actual program statements. The program statements

READ (KARDS,5) NV, NS

5   FORMAT (2I4)

would, if inserted at the appropriate location in our program, cause the computer to read a card on which the user had punched values for NV and NS. It is necessary to adopt some jargon in order to communicate with users concerning the operation of a library program, so we shall refer to a card containing problem parameters as a *parameter card*. We would, of course, provide for input of the parameter card before input of the data cards.

There remains a problem associated with input of the parameter card. What would happen if the user inadvertently supplied a card containing a letter of the alphabet punched in the field from which a value for NV was to be read? The results would vary from computer to computer, but it is likely that the best we could hope for would be a rather technical message like

ILLEGAL CHARACTER ENCOUNTERED IN NUMERIC INPUT STRING

It is probable that the user would not immediately associate this message with a mispunched parameter card. A more explicit error message would be helpful.

We could avoid an obscure computer-system error message by reading the parameter card in A (alphanumeric) conversion. The contents of the parameter card would be read as characters. We could then convert the alphanumeric representation of the contents of the parameter card to numeric representation, taking care to inspect each column for a nonnumeric, nonblank entry.

SUBROUTINE START

This subprogram provides the capability for input of control cards in alphanumeric conversion. The routine is written for the following general control card scheme:

Title card(s) (optional)
Parameter card
Format card(s)
Data cards
Blank card

System control cards ("JOB" card, etc.) have been omitted from the list above. Additional control cards may be added to this scheme as required by specific programs. The calling sequence for subroutine START is

CALL START (KP,KARDS,LINES)

where KP = one-dimensional integer array dimensioned at least 20 in the calling
        program
 KARDS = device number for the card reader
 LINES = device number for the line printer

The variables KARDS and LINES must be defined before subroutine START is called; KP is returned containing the entries punched in 20 four-column fields of the parameter card. The operation of subroutine START is portrayed in the flow chart appearing in Figure 2-3. A listing of the routine appears in Appendix A.

As noted in the flow chart, a *title card* is defined as any card with a nonblank, nonnumeric character punched in column 1. The user may include any

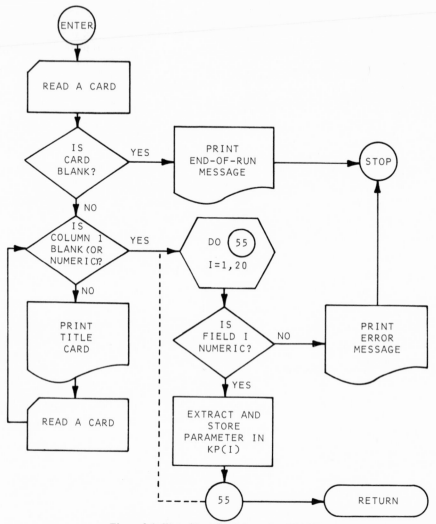

Figure 2-3. Flow Chart for Subroutine START

number of title cards at the beginning of his problem deck, or he may omit the title card entirely. In either case, the first card with a blank or numeral in column 1 is interpreted as the *parameter card*. Subroutine START attempts to extract 20 parameters from 20 four-column fields of the parameter card. Any character other than a blank or numeral punched anywhere on the parameter card results in an error message and termination of execution. Embedded or trailing blanks are interpreted as zero; hence, entries must be right-justified. If the first card by subroutine START is entirely blank, a normal exit occurs.

VARIABLE FORMAT

Although we now have a plan for elimination of the arithmetic statements defining NV and NS in Version II, the 5 FORMAT statement remains. We need some way to input format specifications at execution time rather than as part of the program itself, as in Version II. On many computers, FORTRAN provides us with this capability. Consider the following format specifications:

$$(5X, 10F2.0)$$

Let us assume that these specifications (identical to those in the 5 FORMAT statement of Version II) are punched on a tab card with the 5 FORMAT omitted. The sequence of statements

```
      READ (KARDS,10) (KF(I), I = 1,20)
   10 FORMAT (20A4)
      DO 15 I = 1,NS
      READ (KARDS,KF) (X(J), J = 1,NV)
      DO 15 J = 1,NV
   15 SUMX(J) = SUMX(J) + X(J)
```

would cause the card containing the format specifications to be read according to the 10 FORMAT statement. These specifications are then utilized in the READ statement inside the DO 15 loop. Instead of a reference to a FORMAT statement number, the READ statement refers directly to the name of the array containing the format specifications. In summary, then, format specifications enclosed in parentheses are placed in an array, and the name of the array (without subscripts) is referred to in a READ statement as if the array name were a FORMAT statement number. Any standard format specifications (slashes, A, I, E, X, etc.) may appear within the parentheses, although, of course, the usual rules concerning input-output list and format correspondence apply.

SUBROUTINE MPRT

The sequence of statements associated with output of the means computed in Version II can be replaced by a single call to subroutine MPRT, a routine that prints a one- or two-dimensional real array in eight-element segments or eight-column blocks, respectively. The calling sequence is

CALL MPRT (X,NR,NC,LINES,ND)

where X is the one- or two-dimensional real array to be printed. If X is a two-dimensional array, NR is the number of rows of X to be printed, NC is the number of columns of X to be printed, and ND is the number of rows dimensioned for X in the calling program. If X is a one-dimensional array, NR is the number of elements of X to be printed, NC must be defined as 1, and ND is the number of elements dimensioned for X. The device number for the line printer is LINES. All arguments of subroutine MPRT must be defined before the routine is called. It is important to note that an array may be dimensioned larger than the area actually printed. For example, the statement

CALL MPRT (Y,20,45,5,40)

would cause a 20 by 45 area of the array Y dimensioned 40 by at least 45 to be printed on device 5, and the statement

CALL MPRT (Z,20,1,6,100)

would cause the first 20 elements of an array Z dimensioned 100 to be printed on device 6. A listing of subroutine MPRT appears in Appendix A.

Version III

In addition to the utilization of variable format, subroutine START and subroutine MPRT, this version of our program to compute means incorporates the printing of a program identification, echo checks, and error messages. Figure 2-4 contains a listing of Version III, Figure 2-5 contains a listing of a sample data set used to test Version III, and output from Version III appears in Figure 2-6.

PROGRAM IDENTIFICATION

The first printed output of a library program should be a title identifying the name of the program, version date, and computer installation where the program was developed. We shall consider this and other standards for library programs further in Chapter 7.

ECHO CHECKS

After a call to subroutine START, the first three elements of KP contain the number of variables, number of subjects, and number of format cards to be read, respectively. Accordingly, the variables NV, NS, and NFC are defined, and the values entered are printed along with a listing of the parameter card columns in which they were entered and their meanings. This simple echoing of the user's entries serves at least two purposes. First, it provides documentation of the problem parameters on the program's output for later reference. Second, although the user might be assumed to be fully aware of the values and meanings of the parameters he has entered for a given run, it is possible he has misunder-

```
C
C      ROUTINE TO COMPUTE MEANS - VERSION III
C
       DIMENSION SUMX(100), X(100), KP(20), KF(180)
C
C      DEFINE I/O DEVICE NUMBERS.
C
       KARDS = 2
       LINES = 5
C
C      PRINT PROGRAM IDENTIFICATION.
C
     5 WRITE (LINES,10)
    10 FORMAT (44H1PROGRAM TO COMPUTE MEANS, VERSION OF 1/1/75      /
      1 59H ACADEMIC UNIVERSITY COMPUTER CENTER, FICTIONVILLE, INDIANA /)
C
C      INPUT TITLE AND PARAMETER CARDS.
C
       CALL START (KP,KARDS,LINES)
C
C      NV = NUMBER OF VARIABLES.
C      NS = NUMBER OF SUBJECTS.
C      NFC = NUMBER OF FORMAT CARDS.
C
       NV = KP(1)
       NS = KP(2)
       NFC = KP(3)
C
C      PRINT ECHO CHECKS.
C
       WRITE (LINES,15) NV, NS, NFC
    15 FORMAT (23HOPARAMETER CARD ENTRIES /
      1 24H COLUMNS  ENTRY  MEANING / 6HO  1-4, I8, 3X,
      2 19HNUMBER OF VARIABLES / 6H    5-8, I8, 3X, 18HNUMBER OF SUBJECTS
      3 / 7H   9-12, I7, 3X, 22HNUMBER OF FORMAT CARDS)
C
C      TEST VALIDITY OF PARAMETERS.
C
       IF (NV) 20,20,30
    20 WRITE (LINES,25)
    25 FORMAT (26HOZERO VARIABLES SPECIFIED.)
       CALL EXIT
    30 IF (NV - 100) 45,45,35
    35 WRITE (LINES,40)
    40 FORMAT (32HOTOO MANY VARIABLES (MAX = 100).)
       CALL EXIT
    45 IF (NS) 50,50,60
    50 WRITE (LINES,55)
    55 FORMAT (25HOZERO SUBJECTS SPECIFIED.)
       CALL EXIT
    60 IF (NFC) 65,65,75
    65 WRITE (LINES,70)
    70 FORMAT (29HOZERO FORMAT CARDS SPECIFIED.)
       CALL EXIT
    75 IF (NFC - 9) 90,90,80
    80 WRITE (LINES,85)
    85 FORMAT (33HOTOO MANY FORMAT CARDS (MAX = 9).)
       CALL EXIT
C
C      INPUT AND PRINT FORMAT CARD(S).
C
    90 NT = NFC * 20
       READ (KARDS,95) (KF(I), I = 1,NT)
    95 FORMAT (20A4)
       WRITE (LINES,100) (KF(I), I = 1,NT)
   100 FORMAT (18HOFORMAT CARD(S) = // 9(1X, 20A4))
C
C      INITIALIZE SUMX.
C
       DO 105 I = 1,NV
   105 SUMX(I) = 0.0
C
C      INPUT DATA FOR NS SUBJECTS AND ACCUMULATE IN SUMX.
C
       DO 110 I = 1,NS
       READ (KARDS,KF) (X(J), J = 1,NV)
       DO 110 J = 1,NV
   110 SUMX(J) = SUMX(J) + X(J)
C
C      COMPUTE MEANS.
C
       SN = NS
       DO 120 J = 1,NV
   120 X(J) = SUMX(J) / SN
C
C      PRINT RESULTS AND RETURN TO START.
C
       WRITE (LINES,125)
   125 FORMAT (6HOMEANS)
       CALL MPRT (X,NV,1,LINES,100)
       GO TO 5
       END
```

Figure 2-4. Routine to Compute Means, Version III

```
SAMPLE DATA SET ONE
A TEST OF ROUTINE TO COMPUTE MEANS
    5  12   1
(5X, 5F2.0)
       1  2 612  1  3
       2  3 613  2  1
       3  1 314  1  7
       4  4 116  1  3
       5  2 512  1  3
       6  2 911  2  2
       7  1  8  8  1  1
       8  3 213  1  3
       9  2 114  1  4
      10  1  7  6  2  2
      11  1 612  1  1
      12  1  3  8  1  8
SAMPLE DATA SET TWO
    3  10   1
(5X, 3F2.0)
       1  1  8  6
       2  2  8  6
       3  2  7  6
       4  1  8  3
       5  3  6  3
       6  2  4  6
       7  1  8  4
       8  6  3  5
       9  3  9  5
      10  2  2  4
(BLANK CARD)
```

Figure 2-5. Sample Data Set 1

stood the program documentation or made an error in keypunching his control cards. In either event, echo checks can provide useful information in a post-mortem examination of computer output after a run resulting in abnormal termination or obviously faulty output. If, for example, the user failed to right-justify his parameter card entries, the program might print 250 for the number of variables instead of the intended 25.

ERROR MESSAGES

A substantial portion of the program is devoted to tests of the validity of parameter card entries. Although we are, of course, unable to ascertain whether the user really meant to enter 25 variables instead of 52, we *can* catch certain obvious errors and provide error messages pinpointing the source of difficulty. The devotion of highly compulsive attention to the inclusion of extensive error checks and associated messages can reduce the need for access to supporting personnel by the user. The user who is able to identify and correct his own errors will rejoice in his newly found independence and the attendant increase in his speed and efficiency as a data processor.

MULTIPLE FORMAT CARDS

Occasionally a data deck will require an inordinately complex set of format specifications exceeding the capacity of a single card; hence, it is desirable to provide an option for multiple format cards. Since each format card is read in our program by

```
PROGRAM TO COMPUTE MEANS, VERSION OF 1/1/75
ACADEMIC UNIVERSITY COMPUTER CENTER, FICTIONVILLE, INDIANA

SAMPLE DATA SET ONE
A TEST OF ROUTINE TO COMPUTE MEANS

PARAMETER CARD ENTRIES
COLUMNS  ENTRY  MEANING

  1-4      5    NUMBER OF VARIABLES
  5-8     12    NUMBER OF SUBJECTS
  9-12     1    NUMBER OF FORMAT CARDS

FORMAT CARD(S) =

(5X, 5F2.0)


MEANS

       1        2        3        4        5
     1.916    4.750   11.583    1.250    3.166




PROGRAM TO COMPUTE MEANS, VERSION OF 1/1/75
ACADEMIC UNIVERSITY COMPUTER CENTER, FICTIONVILLE, INDIANA

SAMPLE DATA SET TWO

PARAMETER CARD ENTRIES
COLUMNS  ENTRY  MEANING

  1-4      3    NUMBER OF VARIABLES
  5-8     10    NUMBER OF SUBJECTS
  9-12     1    NUMBER OF FORMAT CARDS

FORMAT CARD(S) =

(5X, 3F2.0)


MEANS

       1        2        3
     2.300    6.300    4.800




PROGRAM TO COMPUTE MEANS, VERSION OF 1/1/75
ACADEMIC UNIVERSITY COMPUTER CENTER, FICTIONVILLE, INDIANA

END OF JOB -- BLANK CARD ENCOUNTERED.
```

Figure 2-6. Output from Version III Program to Compute Means

95 FORMAT (20A4)

requiring one array element for each four format-card columns, input of up to nine format cards requires the array KF to be dimensioned 180. Computers vary in the number of characters that can be stored in a single integer-array element; variations in this capacity would require associated changes in the program statements related to input of the variable format specifications.

USE OF BLANK CONTROL CARD

The statement

GO TO 5

just before the END statement causes an additional call to subroutine START, giving the user the option to "stack" data sets for processing. If a given data set is followed by a blank card rather than a title or parameter card for an additional data set, a normal program termination occurs.

The sample data set appearing in Figure 2-5 includes only control cards and data processed by Version III; system control cards are excluded. The corresponding computer output appears in Figure 2-6.

At this stage in its development, the routine to compute means is a full-fledged library program. On many computer systems it would be possible to store Version III in a form suitable for immediate execution, eliminating the overhead associated with compilation and loading. In this case, a single system control card might suffice to "invoke" our program and turn control of processing over to it. The user would never see a source or object deck, and he would therefore have no opportunity to shuffle program deck card order.

Simulating Variable Format

The form of variable format (also called run-time format specification or object-time format specification) presented in connection with Version III is not available on all computers. Various simulations of this feature have been devised, and the purpose of this section is to present one such simulation.

Although several small-scale computing systems such as the IBM 1130 or the Digital Equipment Corporation PDP-11 feature versions of FORTRAN in which "true" variable format is not available, it is nevertheless possible to simulate the essential characteristics of variable format with FORTRAN subroutines. The use of subroutine START and the subroutines described below eliminate the need for the statistical library program user to modify actual source statements; accordingly, precompiled library programs may be stored on system mass storage and called up for execution by the user.

CHARACTERISTICS OF FORMAT SPECIFICATIONS

From the point of view of the library program user, the simulated variable format described below resembles very closely "full-scale" or "real" variable format with the following restrictions:

1. Only X and F field specifications are permitted.
2. Only one level of nested parentheses is permitted.

Within the above restrictions, the following operating characteristics apply:

1. All blanks within the format specifications are ignored.
2. Redundant parentheses are ignored.

3. Redundant commas are ignored.

4. The specification X is taken as equivalent to 1X.

5. The comma following X may be omitted.

6. In a field specification of the form rFw.d, the value of "r," if omitted, is taken as 1.

7. In a field specification of the form rFw.d, the value of "d" may be greater than the value of "w" (e.g., F3.4 is permitted).

8. Slashes are permitted.

The following are illustrations of permissible format specifications:

$$(5X, 10F5.0, 5F5.4)$$
$$(2X,3F2.0/5X,F4.1,F2.3)$$
$$(2(5X, 3F2.0, X, F1.0), 2XF1.0)$$
$$(5X,,,2F1.0 //)$$
$$((F4.07))$$

The following are examples of invalid format specifications:

Specifications	Error
(5X, 10I1)	Specification I not permitted
(5(3(2X,F1.0)))	Excessive nesting
(A4, 10F2.0)	Specification A not permitted

In addition to the above features, the user has the option of continuing his format specifications on a second card by punching an asterisk (*) in column 80 of his first format card.

DATA INPUT CHARACTERISTICS

Several features of the variable format simulation with respect to input data fields should be noted.

1. If the user-supplied format specifications do not satisfy the input list of the program, the format specifications will be rescanned in an effort to satisfy the input list.

2. Although minus (-) signs are permissible, no plus (+) signs are permitted.

3. A decimal point in the data field overrides the corresponding format specification.

4. E = format data are not permitted.

5. Embedded and trailing blanks are converted as zeros.

The following examples illustrate the operation of the simulated variable format during data input (b = blank):

Specification	Field	Result
F3.0	bb7	7.0
F3.4	bb7	0.0007
F2.1	b7	0.7
F3.0	b+7	Error (+ not permitted)
F3.0	–b7	– 7.0
F3.0	–7b	–70.0
F5.2	b6b3b	60.30

USE IN PROGRAMS

Two subroutines are used in this simulation of variable format: subroutine FMAT for input and decoding of the format specifications and subroutine TAKE for input of data according to the format specifications.

SUBROUTINE FMAT. This subroutine inputs the format card(s), echos the format card(s), scans the format specifications for errors, and stores a coded form of the format specifications in an array. The calling sequence is

CALL FMAT (KF,KARDS,LINES,LIMIT)

where   KF = one-dimensional integer array
        KARDS = device number for the card reader
        LINES = device number for the line printer
        LIMIT = number of F fields and slashes to be permitted

Since three elements of KF are required for each F field or slash, KF must be dimensioned at least 3 * LIMIT in the calling program. The following examples illustrate the storage requirements of typical format specifications:

Format	Elements of KF
(5X, F3.0)	3
(2(5X, F3.0))	6
(3(5X, 4F1.0))	36
(/ 10F2.0)	33
(5X, 4(3X2F1.0)/)	27

In practice, the value of LIMIT may be established as $S + 10$, where S is the maximum number of scores the program user would ever attempt to read for each entity (subject). Accordingly, a program to compute intercorrelations among 40 variables would require values of 50 and 150 for LIMIT and the dimension of KF, respectively, and a program to perform a groups-by-trials analysis of variance (i.e., two-factor design with repeated measures on one factor) for up to 10 groups, 5 trials (repeated measures), and 20 dependent variables would require values of 110 and 330 for LIMIT and the dimension of KF, respectively.

If subroutine FMAT encounters an error among the format specifications, it prints an error message and terminates execution.

SUBROUTINE TAKE. Input of each entity's (subject's) data is accomplished by a call to subroutine TAKE. The calling sequence is

CALL TAKE (KF,X,NV,KARDS,LINES)

where

KF = one-dimensional integer array containing the format specifications encoded by subroutine FMAT

X = one-dimensional real array in which scores are to be placed, beginning with X(1)

NV = number of scores to be read and placed in X

KARDS = device number for the card reader

LINES = device number for the line printer

The call to subroutine TAKE would usually appear in a DO loop indexed from 1 to the number of entities (subjects).

Version IV

This version of the routine to compute means (Figure 2-7) utilizes the simulation of variable format presented above. Because subroutine FMAT decides whether to read one or two format cards, the parameter card entry specifying the number of format cards to be read has been omitted. As in Version III, the array KF has been used to store format specifications; however, the dimension of KF has been increased to meet the requirements of subroutine FMAT. As can be seen in statement 60, the call to subroutine FMAT includes a value of 110 for the argument LIMIT (a maximum of 100 scores per subject + 10). Accordingly, KF has been dimensioned 330 (3 * LIMIT).

The call to subroutine TAKE in the DO 70 loop indexed from 1 to NS (the number of subjects) causes all the scores for one subject to be read in each cycle of the loop.

The same sample data set used to test Version III could be used to test Version IV, and the output from the two versions would differ only with respect to the echo checks of the parameter card entries (columns 9 through 12 are unused in the case of Version IV).

```
C
C     ROUTINE TO COMPUTE MEANS - VERSION IV
C
      DIMENSION SUMX(100), X(100), KP(20), KF(330)
C
C     DEFINE I/O DEVICE NUMBERS.
C
      KARDS = 2
      LINES = 5
C
C     PRINT PROGRAM IDENTIFICATION.
C
    5 WRITE (LINES,10)
   10 FORMAT (44H1PROGRAM TO COMPUTE MEANS, VERSION OF 1/1/75     /
     1 59H ACADEMIC UNIVERSITY COMPUTER CENTER, FICTIONVILLE, INDIANA /)
C
C     INPUT TITLE AND PARAMETER CARDS.
C
      CALL START (KP,KARDS,LINES)
C
C     NV = NUMBER OF VARIABLES.
C     NS = NUMBER OF SUBJECTS.
C
      NV = KP(1)
      NS = KP(2)
C
C     PRINT ECHO CHECKS.
C
      WRITE (LINES,15) NV, NS
   15 FORMAT (23H0PARAMETER CARD ENTRIES /
     1 24H COLUMNS   ENTRY   MEANING / 6H0   1-4, I8, 3X,
     2 19HNUMBER OF VARIABLES / 6H   5-8, I8, 3X, 18HNUMBER OF SUBJECTS)
C
C     TEST VALIDITY OF PARAMETERS.
C
      IF (NV) 20,20,30
   20 WRITE (LINES,25)
   25 FORMAT (26H0ZERO VARIABLES SPECIFIED.)
      CALL EXIT
   30 IF (NV - 100) 45,45,35
   35 WRITE (LINES,40)
   40 FORMAT (32H0TOO MANY VARIABLES (MAX = 100).)
      CALL EXIT
   45 IF (NS) 50,50,60
   50 WRITE (LINES,55)
   55 FORMAT (25H0ZERO SUBJECTS SPECIFIED.)
      CALL EXIT
C
C     INPUT AND PRINT FORMAT CARD(S).
C
   60 CALL FMAT (KF,KARDS,LINES,110)
C
C     INITIALIZE SUMX.
C
      DO 65 I = 1,NV
   65 SUMX(I) = 0.0
C
C     INPUT DATA FOR NS SUBJECTS AND ACCUMLLATE IN SUMX.
C
      DO 70 I = 1,NS
      CALL TAKE (KF,X,NV,KARDS,LINES)
      DO 70 J = 1,NV
   70 SUMX(J) = SUMX(J) + X(J)
C
C     COMPUTE MEANS.
C
      SN = NS
      DO 75 J = 1,NV
   75 X(J) = SUMX(J) / SN
C
C     PRINT RESULTS AND RETURN TO START.
C
      WRITE (LINES,80)
   80 FORMAT (6H0MEANS)
      CALL MPRT (X,NV,1,LINES,100)
      GO TO 5
      END
```

Figure 2-7. Routine to Compute Means, Version IV

Additional Control Cards

Earlier we presented a general control scheme consisting of the card order

1. Title card(s) (optional)
2. Parameter card
3. Format card(s)
4. Data cards
5. Blank card

This basic scheme, implemented in Versions III and IV, may be augmented by such additional control cards as may be required by specific programs. In this section we consider two examples of additional control cards.

VARIABLE NAMES

Identification of variables on computer output has so far been limited to assignment of an integer corresponding to serial order of input. This information, used in conjunction with title and format card entries, should suffice for positive identification of input variables, but it is convenient to have appropriate names printed on the computer output by the program itself. Accordingly, let us augment our control card list to read

1. Title card(s) (optional)
2. Parameter card
3. Format card(s)
4. Variable names card(s) (optional)
5. Data cards
6. Blank card

An additional parameter card field is allocated for exercise of the variable names option. If the field is left blank, no variable names cards are expected; a 1 in the field causes variable names to be read. The variable names may be stored in a two-dimensional array NAMES:

READ (KARDS,50) ((NAMES(I,J), J = 1,10), I = 1,NV)

50   FORMAT (40A2)

The array NAMES is dimensioned NV by 10; that is, NAMES has as many rows as the maximum numbet of variables permitted by the program. Each variable name may consist of up to 20 characters (including blanks), and variable names are entered left-justified in 20-column fields on the variable names cards in the same order as that for input of the variables.

The storage of only two characters in each integer array element represents an effort to maintain compatibility with such small computers as the IBM 1130; the dimensions of NAMES and the associated program statements could be easily modified for a machine permitting the storage of more characters in a single integer-array element.

The output sequence of our routine to compute means is modified to include two output sections, one using subroutine MPRT without variable names, and one consisting of the following statements:

130   WRITE (LINES,135)
135   FORMAT (49H0VARIABLE   DESCRIPTION   MEAN/)
      DO 140 I = 1,NV

```
140    WRITE (LINES,145) I, (NAMES (I,J), J = 1,10), X(I)
145    FORMAT (I6, 13X, 10A2, F11.3)
```

The appropriate parameter card entry is tested in an IF statement to determine which output statements to execute.

If core restrictions pose a serious problem, variable names may be stored on magnetic tape or disk. The following sequence of statements causes variable names to be stored on scratch tape NTAPE:

```
        REWIND NTAPE
        DO 100 I = 1,NV,4
        READ (KARDS,95) (NAMES(J), J = 1,40)
95      FORMAT (40A2)
        DO 100 J = 1,40,10
        JJ = J + 9
100     WRITE (NTAPE) (NAMES(K), K = J,JJ)
```

In this case, NAMES is dimensioned only 40. The corresponding output statements for our routine to compute means are as follows:

```
        REWIND NTAPE
130     WRITE (LINES,135)
135     FORMAT (49H0VARIABLE    DESCRIPTION    MEAN /)
        DO 140 I = 1,NV
        READ (NTAPE) (NAMES(J), J = 1,10)
140     WRITE (LINES,145) I, (NAMES(J), J = 1, 10), X(I)
145     FORMAT (I6, 13X, 10A2, F11.3)
```

The printing of a two-dimensional array with associated variable names may be accomplished using subroutine MPRTN presented in Appendix A.

MISSING DATA AND SUBROUTINE MISS

Quite often subjects in an experiment or research project have missed certain testings or denied the researcher his data in some other way. As a result, instances of missing data will be sprinkled throughout a data file. The program library user possessing such incomplete data will often desire to run analyses in which only subjects with complete data on a given variable are included in the analysis for that variable. For example, in our routine to compute means, we might wish the mean for a given variable to be based only upon data from subjects with a valid score.

Common practice is to let zero represent an instance of missing data. This means that care must be taken to avoid the occurrence of zero as a valid score. The application of a linear transformation to the raw data can eliminate such difficulties. For example, raw scores on an achievement test might range from 0 to 50. The addition of 1 to each subject's raw score would result in transformed scores ranging from 1 to 51. Subtraction of 1 from the mean would convert the mean to the original metric; the variance would, of course, be unaffected.

Provision of an option of inspection for missing data occasions an additional control card:

1. Title card(s) (optional)
2. Parameter card
3. Format card(s)
4. Variable names card(s) (optional)
5. Missing data signal card(s) (optional)
6. Data cards
7. Blank card

Yet another parameter card field is utilized to signal the computer to read one or more missing data signal cards. If there are 80 or fewer variables, each column of a single missing data signal card corresponds to a variable; otherwise, column 1 of the second missing data signal card corresponds to variable 81, etc. An entry of zero in a column of the missing data signal card signals that zero scores and blank fields for the corresponding variable are to be treated as valid scores, just as 1 signals that zero scores and blank fields are to signify missing data.

Subroutine MISS may be used to input and echo a single missing data signal card (corresponding to 80 or fewer variables). The calling sequence is

CALL MISS (M,NV,KARDS,LINES)

where $\qquad$ M = one-dimensional integer vector (dimensioned with at least NV elements) returned containing NV ones and zeros

NV = number of variables

KARDS = device number for the card reader

LINES = device number for the line printer

Because subroutine MISS will not input more than a single missing data signal card, some other method of input must be found if NV is greater than 80. A listing of subroutine MISS is included in Appendix A.

Input of data and computation of means are modified as follows:

```
        DO 115 I = 1,NV
        SUMX(I) = 0.0
115     NVS(I) = NS
        DO 120 I = 1,NS
        READ (KARDS,KF) (X(J), J = 1,NV)
        DO 120 J = 1,NV
```

```
          IF (M(J)) 120,120,110
   110    IF (X(J)) 120,115,120
   115    NVS(J) = NVS(J) -1
   120    SUMX(J) = SUMX(J) + X(J)
   C      COMPUTE MEANS.
          DO 130 I = 1,NV
          IF (NVS(I)) 130,130,125
   125    X(J) = SUMX(J) / FLOAT(NVS(J))
   130    CONTINUE
```

The output is modified to include for each variable the number of subjects upon which the mean is based. Also, if NVS(I) = 0 for the ith variable, then an appropriate message is printed rather than the mean.

The control cards introduced in this section are intended to be only suggestive of the ways in which additional control cards may be added to the basic scheme as required by a specific program. The creative programmer will quickly recognize numerous possibilities for increasing the flexibility of a program through the addition of new control cards. A word of caution is in order, however. In Chapter 1 it was mentioned that an increase in the flexibility of a program was generally accompanied by increased complexity from the user's point of view. A point of diminishing returns is quickly reached as the number of control cards increases. It is better to keep a routine relatively simple to use (albeit rather limited in flexibility) than to overwhelm the timorous user with a bristling array of control cards.

ALTERNATIVE CONTROL CARD SYSTEMS

Although we have considered elaborations of our control card system, the basic system remains

1. Title card(s) (optional)
2. Parameter card
3. Format card(s)
4. Data cards
5. Blank card

The feasibility of this system depends upon the availability of some form of variable format. If the FORTRAN available to the programmer does not include variable format and no simulation of the feature is available, an alternative control card scheme will be required. Two such schemes are described below.

The Minimal Program Deck

In this approach the order of input becomes

1. Program deck (including any required system control cards)
2. Title card(s) (optional)
3. Parameter card
4. Data cards
5. Blank card

An example of the program deck appears in Figure 2-8. The sole purpose for the existence of the program deck is to permit the user to change the 3 FORMAT statement in accordance with his data. Since any cards handled by the user are liable to disarrangement, as few cards as possible are included in the program deck. Several structural features aid in the minimization of deck size and general visual confusion of the user.

```
C
C      ROUTINE TC CCMPUTE MEANS - VERSION V
C
       COMMON KARDS, LINES, NV, NS, X(200)
     1 CALL SUB1
       CC 2 I = 1,NS
       READ (KARDS,3) (X(J), J = 1,NV)
     2 CALL SUB2
       CALL SUB3
       GO TO 1
C
C      INSERT ''3 FORMAT'' CARD NEXT.
C
     3 FORMAT (5X, 10F2.0)
       END
```

Figure 2-8. Routine to Compute Means, Version V

USE OF COMMON

Communication with all subroutines called in the program deck is via COMMON. This approach eliminates argument lists associated with the subroutines and helps "streamline" the appearance of the program deck. The COMMON statement itself is shortened somewhat by the declaration of a single 200-element array X instead of X(100), SUMX(100). The COMMON statement included in the subroutines SUB1, SUB2, and SUB3 would be as follows:

COMMON KARDS, LINES, NV, NS, X(100), SUMX(100)

USE OF SUBROUTINES

It is assumed that the subroutines SUB1, SUB2, SUB3, and any other subprograms required for execution of the main program of the program deck would be stored as object code in mass storage (e.g., magnetic disk or tape) associated with the computer system, thus eliminating the need for the user to handle them. Subroutine SUB1 would include definition of input/output device numbers (KARDS and LINES), printing of program identification, a call to subroutine START, printing of echo checks, testing of parameters for validity, and initialization of SUMX (see Figure 2-4). In subroutine SUB2, sums would be accumulated in SUMX, and in subroutine SUB3 means would be computed and results printed. The use of subroutines rather than direct coding of operations within the main program greatly reduces the size of the program deck the user must manipulate.

SEQUENCE NUMBERING. Although not shown in Figure 2-8, it would be highly desirable for the cards in the program deck to include an identification sequence in columns 73 through 80. If system control cards of the computer used also include an identification sequence in these columns, any unchanging system control cards should be included in the numbering scheme.

ADDITIONAL FEATURES. The statement

GO TO 1

permits the user to "stack" data sets for continuous processing, but it is important to note that such stacking involves the use of the same format specifications with each successive data set. While we have used the subroutine names SUB1, SUB2, and SUB3 in our example here, it is advantageous to use subroutine names related to the name given the main program. For example, if the main program were known as X23, suitable subroutine names might be X23A, X23B, etc.

The Variable Location Card

If the input to our library program is to consist entirely of single-column scores (e.g., item responses or demographic information such as sex or religious affiliation), it is possible to use the following control card scheme:

> TITLE CARD(S) (OPTIONAL)
> PARAMETER CARD
> VARIABLE LOCATION CARD(S)
> DATA CARDS
> BLANK CARD

Each of the variable location cards corresponds to one of the data cards in a subject's card set. For example, if the data cards include three cards per subject, three variable location cards would be required.

There is a one-to-one correspondence between variable location card columns and data card columns. Thus, a 1 (one) would be punched in each variable location card column that corresponds to a data card column in which scores for a variable appear. Other variable location card columns would be left blank. The inclusion of a blank card as a variable location card would simply signify that no information was to be read from the corresponding card in each subject's card set.

SUBROUTINE DCODE

Input of the variable location cards involves utilization of an additional subroutine with the following calling sequence:

CALL DCODE (KK,NSTAR,NSTOP,INTEG,IER)

This subroutine converts a string of alphanumeric characters to an integer number. Only positive integers may be extracted, and blanks (leading, trailing, or embedded) are converted as zeros. The characters to be converted must be stored one character per element of a one-dimensional integer array. In the preceding call statement,

KK = integer array

NSTAR = leftmost position (lowest-numbered element) in KK to be decoded

NSTOP = rightmost position (highest-numbered element) in KK to be decoded

INTEG = integer variable that will contain the result decoded

IER = error indicator

The indicator IER is an integer variable that will equal zero upon return if only blanks or numeric characters are encountered in the field. If a nonnumeric, nonblank character is encountered, IER will contain the number of the element of KK in which the error occurred, and INTEG will be defined as zero. Therefore, IER should be tested after each call to subroutine DCODE. A listing of subroutine DCODE appears in Appendix A.

INPUT OF VARIABLE LOCATION CARDS

The following sequence of statements illustrates input and error checks for a set of variable location cards (where NV = number of variables and NCS = number of cards per subject):

```
    NSPEC = 0
    DO 125 I = 1,NCS
    READ (KARDS,100) (KK(J), J = 1,80)
100 FORMAT (80A1)
```

```
        DO 125 J = 1,80
        CALL DCODE (KK,J,J,KF(I,J),IER)
        IF (IER) 105,105,115
105     IF (KF(I,J) - 1) 125,110,115
110     NSPEC = NSPEC + 1
        GO TO 125
115     WRITE (LINES,120) J, (KK(K), K = 1,80)
120     FORMAT (24H0INVALID ENTRY IN COLUMN, 13, 49H OF PRE-
        1SUMED VARIABLE LOCATION CARD ON NEXT LINE. / 1X, 80A1)
        CALL EXIT
125     CONTINUE
        IF (NV - NSPEC) 130,140,130
130     WRITE (LINES,135)
135     FORMAT (74H0NUMBER OF ENTRIES ON VARIABLE LOCATION
        1CARD(S) DOES NOT AGREE WITH NUMBER / 24H OF VARIABLES
        2SPECIFIED.)
        CALL EXIT
140     (next statement in program)
```

After execution of the preceding sequence, the array KF would contain ones and zeros corresponding to the ones and blanks of the variable location cards; hence, KF would be dimensioned 9 by 80 if up to nine cards per subject were permitted. The tests of the inner DO 125 loop demand that the variable location cards contain only blanks, zeros, or ones; the IF statement after statement 125 demands that the number of ones appearing on the variable location cards correspond to the total number of variables specified on the parameter card.

INPUT OF DATA

In the following sequence of statements the information obtained from the variable location cards is utilized during input of data.

```
        DO 225 I = 1,NS
        NVC = 0
        DO 225 J = 1,NCS
        READ (KARDS,200) (KK(K), K = 1,80)
200     FORMAT (80A1)
        DO 225 K = 1,80
        IF (KF(J,K)) 225,225,205
205     NVC = NVC + 1
        CALL DCODE (KK,K,K,KTEMP,IER)
        IF (IER) 220,220,210
```

```
210   WRITE (LINES,215) K, (KK(L), L = 1,80)
215   FORMAT (24H0INVALID ENTRY IN COLUMN, I3, 26H OF DATA
      1CARD ON NEXT LINE / 1X, 80A1)
      CALL EXIT
220   X(NVC) = KTEMP
      .
      .
      .
      (program statements)
      .
      .
      .
225   CONTINUE
```

In this sequence, NS = number of subjects and NCS = number of cards per subject. The array KF is assumed to have been defined during input of the variable location cards, and X is the array in which scores are to be placed for subsequent manipulation. The use of subroutine DCODE permits an explicit response to the occurrence of a nonblank, nonnumeric character in a data field; however, this method of input might prove to be too slow in some situations. If so, the following simplification might be used:

```
      DO 210 I = 1,NS
      NVC = 0
      DO 210 J = 1,NCS
      READ (KARDS,200) (KK(K), K = 1,80)
200   FORMAT (80I1)
      DO 210 K = 1,80
      IF (KF(J,K)) 210,210,205
205   NVC = NVC + 1
      X(NVC) = KK(K)
      .
      .
      .
      (program statements)
      .
      .
      .
210   CONTINUE
```

In this case, a nonblank, nonnumeric character in any field of any data card would result in a fatal (and, depending upon the computer system in use, often rather obscure) error. If subroutine DCODE is used, it is not necessary, of course, to terminate execution upon detection of an input error. One could, for example, simply set the corresponding element of X equal to zero.

ECHO CHECKS OF VARIABLE LOCATION CARDS

If at all possible, it is highly desirable to print echo checks of variable location card entries. Where results are printed for each variable, echo checks can be included in the output:

```
        WRITE (LINES,300)
   300  FORMAT (37H0VARIABLE      CARD     COLUMN     MEAN /)
        NVC = 0
        DO 315 I = 1,NCS
        DO 315 J = 1,80
        IF (KF(I,J)) 315,315,305
   305  NVC = NVC + 1
        WRITE (LINES,310) NVC, I, J, XBAR(NVC)
   310  FORMAT (I6, 2I9, F14.3)
   315  CONTINUE
```

Other applications and output layouts may require considerable ingenuity on the part of the programmer.

OTHER JARGON

The term "variable location card" is, of course, arbitrary. If the single-column data fields intended for input happened to be items of a test, the user of the program would probably find the term "item location card" more sensible than the term "variable location card." The selection of names for control cards should be guided by careful consideration of the probable significance of the names for library program users.

SUBJECT IDENTIFICATION FIELDS

Because certain statistical procedures (e.g., calculation of predicted criterion scores in regression analysis) require printing of information on a subject-by-subject basis, input of each subject's identification field is occasionally required. This input may be handled in several ways, depending upon the control card scheme in use and the FORTRAN features available.

Use of A-Conversion

When "real" (not simulated variable format) is available, the user can simply be required to include some standard number of A fields at the beginning of his format specifications. If, for example, up to 16 characters could be stored in a double-precision real variable (as in the case of such machines as the IBM 360), a nonsubscripted variable SUBID could be declared double precision, and the statement

$$\text{READ (KARDS,KF) SUBID, (X(J), J = 1,NV)}$$

could be used in the data input loop. The format specifications stored in the one-dimensional array KF would feature a required A field of up to A16, as in the following examples:

(A8, 4X, 20F2.0)
(/A12, 10F3.1)
(A12, 20F1.0 / 12X, 30F1.0)

If, on the other hand, the FORTRAN available on a computer such as the Hewlett-Packard 2100 series permits the use of A-conversion only with integer mode, a one-dimensional array ID could be dimensioned and referenced in the statement for input of each subject's data:

$$\text{READ (KARDS,KF) (ID(J), J = 1,NID), (X(K), K = 1,NV)}$$

The variable NID, the number of characters in each subject's identification field, would be defined by a field on the parameter card. The format specifications stored in KF would begin with NID A1 fields:

(12A1, 20F1.0)
(5X, 6A1, 15F2.0)
(6A1, 2X, 6A1, 10F3.0)

As can be seen in the third example, the subject identification field could consist of two or more noncontiguous sets of characters.

After either of the above READ statements, the subject identification information and scores would typically be stored on magnetic tape or disk for later retrieval and use.

Variable Location Cards and Subject Identification Fields

If variable location cards are to be used, two fields of the parameter card can be used to define the left- and rightmost columns (NL and NR, respectively) of the subject's identification field. Each subject's identification field can then be stored in the one-dimensional integer array ID during input of data:

```
        DO 245 I = 1,NS
        NVC = 0
        DO 240 J = 1,NCS
        READ (KARDS,200) (KK(K), K = 1,80)
200     FORMAT (80A1)
        IF (J - 1) 215,205,215
205     NE = 0
        DO 210 K = NL,NR
        NE = NE + 1
210     ID(NE) = KK(K)
215     DO 240 K = 1,80
        IF (KF(J,K)) 240,240,220
220     NVC = NVC + 1
        CALL DCODE (KK,K,K,KTEMP,IER)      /
        IF (IER) 235,235,225
225     WRITE (LINES,230) K, (KK(L), L = 1,80)
230     FORMAT (24H0INVALID ENTRY IN COLUMN, I3, 26H OF DATA
       1CARD ON NEXT LINE / 1X, 80A1)
        CALL EXIT
235     X(NVC) = KTEMP
240     CONTINUE
        (statements to store ID and data for Ith subject)
245     CONTINUE
```

The array KF contains the variable location card specifications for the NCS cards for each of the NS subjects.

After the earlier call to subroutine START, it would have been highly desirable to make sure that the user's specifications for NL and NR were plausible to the extent that NL > 0, NR > 0, and NR – NL + 1 $\leqslant$ dimension of ID.

Simulated Variable Format and Subject Identification Fields

Again using two fields of the parameter card to define NL and NR as above, a special version of subroutine TAKE is called:

CALL TAKEI (KF,X,NV,ID,NL,NR,KARDS,LINES)

where   KF = one-dimensional integer array containing format specifications encoded by subroutine FMAT

   X = one-dimensional real array in which scor   to be placed, beginning with X(1)

NV = number of scores to be read and placed in X

ID = one-dimensional integer array in which the subject's identification field is to be placed, one character per element

NL = leftmost column of the subject's identification field

NR = rightmost column of the subject's identification field

KARDS = device number for the card reader

LINES = device number for the line printer

The subject's identification field must appear on the first card read for that subject. As in previous methods of reading each subject's identification field, it is assumed that magnetic tape or disk is used during input to store required information for each subject.

SUMMARY

This chapter has presented a basic control scheme suitable for use with versions of FORTRAN featuring variable format. In addition to the basic scheme, it has presented two illustrations of additional control cards that could be used in the implementation of such program features as variable names and inspection for missing data.

This basic control card scheme is only one of many workable approaches. Another may be more appealing to a given programmer; if so, he should use it. The choice of a particular approach to control cards is not so important as the maintenance of consistency among the various programs of a statistical library. It is this consistency that permits the user of a program library to move among the routines of the library with maximal ease and comfort.

The two alternative control card schemes for use without variable format are also intended to be illustrative rather than exhaustive. If they serve only to motivate the creative programmer to devise other solutions for operation without variable format, their presentation will have been useful.

3 Documentation

When a researcher makes the comment that the computer center at Termite Tech has a good statistical program library, it is quite likely that he is reacting to the quality of documentation available to users of the library. Solid documentation is the hallmark of a carefully developed program library; inadequate documentation renders the most elegant of programs utterly useless.

ASSUMPTIONS ABOUT USERS

Before the vital interface between user and program can be constructed, it is essential to analyze carefully the consumers for whom the program is intended. In general, it is better to assume minimal sophistication than to overestimate the typical user's coping ability.

Prior Computer-Related Experience

Assuming that some sort of simple-minded orientation to the tab card and the card punch is available, it is probably reasonable to proceed on the assumption that the user of the statistical library will be familiar with the 80-column tab card and the associated card punch machine. Training of the user in these first few fundamental data processing concepts will almost certainly require some one-to-one assistance and "hands on" experience with the actual equipment involved. While computer center staff will probably have to play a major part in training during the early days of the introduction of computing to a given organization, a "pool of expertise" quickly develops and informal training mechanisms begin to function.

Attitudinal Predisposition

The thought of working with a computer is quite threatening to a sizable proportion of potential users, and care must be taken to create a highly supportive consultative environment in which emphasis is placed upon making the user's first experience with the computer a positive one. If the goal of the staff responsible for support of the statistical library is to make the user self-sufficient as soon as possible, the extra amount of time and effort invested in the beginning user will pay off. The temptation to perform the analysis for the user without requiring him to acquire skills for himself must be resisted.

Statistical Competency

Many naive computer users are unable to separate "statistical consultation" from "data processing consultation." A visit to the computer consultant by a user desiring to learn how to "run my data through the computer" will probably reveal an individual in quest of fertilizer for his parameter farm. Although realization that many statistical library users may require extensive statistical consultation has direct implications for computer center staff requirements, it is impractical to make the statistical library user's manual a statistics textbook. Accordingly, assumption of at least minimal familiarity with the statistical techniques represented by the library programs is justifiable. It is, of course, highly desirable to avoid obscure jargon as much as possible and to point the user to readily accessible and understandable references.

FUNCTIONS OF DOCUMENTATION

In general, a program write-up should perform those functions that would otherwise be handled by consultative staff on a one-to-one basis. Although each program's write-up must clarify any unique attributes of that particular program, it is nevertheless possible to point to several general functions of program documentation.

Quick Familiarization

While each library program is written to handle a relatively large class of analyses, the program is applicable, of course, to only certain sets of data. The user shopping for the program appropriate for his data should be able to ascertain very quickly whether a given program fills his needs. The first part of any program write-up must include a brief overview of the general characteristics of the program, the statistical technique it implements, and the sets of data it can accommodate.

Card Order

Because the user will be using the write-up like a recipe and will want to double-check his problem deck before submitting it for a run, the write-up must provide a list of the control cards in the appropriate order.

Specific Card Preparation

Although it is highly desirable to maintain as much similarity as possible among the control card requirements of the various programs in a statistical library, at least a few programs will require unique control cards. The program write-up must provide, of course, detailed instructions for the preparation of any control cards not described in the introductory sections of the program library user's manual. The choice of computer-related jargon used in describing control cards should, as noted in Chapter 2, be married to the statistical techniques involved.

Background References

As mentioned earlier, it is essential that the program write-up provide the user readable references in the event he should discover that his statistical background is inadequate to permit him to comfortably use the program. If one of several alternative (and somewhat disputed) computational procedures has been adopted, the write-up should contain a note to that effect.

Sample Problem and Output

This section of the write-up can serve to augment the instructions for preparation of specific control cards by providing a concrete example for the user to study. Also, a sample problem provides individuals at other computational facilities a convenient way to test the program after they have attempted to adapt it for use at their installation.

EXAMPLES OF DOCUMENTATION

Although the details of documentation with respect to programs in use at a particular computer installation are obviously determined by the characteristics of the users and the computer system peculiar to that installation, one starting point for the development of documentation is to examine program write-ups from an existing computer center.* The illustrations in this chapter are based on the *Peabody Statistical Library User's Manual*,† which describes a set of programs in use at George Peabody College for Teachers, Nashville, Tennessee.

Introductory Sections

The first three illustrations of this chapter contain the introductory sections of the *Peabody Statistical Library User's Manual*. It is assumed that any user of the Peabody Statistical Library (PSL) will read these sections before he attempts to run a program. The introductory sections were written with the goal of providing sufficiently detailed instructions to permit the careful reader to begin using the PSL without further assistance; however, in practice, the user with no prior computer-related experience usually requires some hand holding during his first contact with the computer center.

INTRODUCTION

This section should sketch the general philosophy behind the structure of the program library and establish the context within which the programs have been written. In Peabody's case, as displayed below, mention is made of the existence of computational facilities at Vanderbilt University, across the street from Peabody.

---

*In fact, the form of the documentation presented in this chapter was inspired by the manual for Biomedical Computer Programs, W. J. Dixon, ed., *Biomedical Computer Programs*, Los Angeles: University of California Press, 1967.

†J. H. Hogge, ed., *Peabody Statistical Library User's Manual*, Nashville, Tenn.: George Peabody College, 1972; J. H. Hogge and Judith Picklesimer, *Peabody Statistical Library User's Manual*, Nashville, Tenn.: George Peabody College, 1971.

PSL USER'S MANUAL

1. Introduction

The Peabody Statistical Library (PSL) is a collection of computer programs designed to provide computational support and data manipulation capabilities for statistical analysis. The programs of the PSL have been devised to provide flexibility of input data layout, simplicity of control cards and reasonable operating efficiency for the IBM 1130 computing system.

Because the Peabody Computer Center is operated on an "open shop" basis, PSL users are expected to run their own analyses. The operation of the IBM 1130 is very straightforward, however, and the Computer Center staff are available to provide the approximately fifteen minutes of training required to become familiar with the system.

The scale of the analyses possible with the PSL is limited primarily by consideration of execution time. Peabody Computer Center users are fortunate to have access to the large computer systems at Vanderbilt University; hence, the user with a statistical analysis impractically large for the 1130 may use the greater computational power of the XDS Sigma 7 at the Vanderbilt Computer Center.

Because descriptions of general Computer Center policy and hardware are provided in another document, the *Peabody Computer Center User's Manual* (presented in Appendix B), such considerations have been excluded from this section of the *PSL User's Manual.*

DATA PREPARATION

Although this section could be expanded to include considerably more material than the example given below, its content is intended to provide at least a general orientation concerning the input characteristics of routines in the program library.

PSL USER'S MANUAL

2. Preparation of Data for the PSL

It is assumed that the user is acquainted with the card punch machine (keypunch) and that he can arrange for the posting of his data to tab cards. The Computer Center offers no keypunch services; however, the user may arrange for such services at commercial data processing organizations in the Nashville area. The Computer Center also maintains a list of individuals who have indicated an interest in providing such services. The discussion of the present section is restricted to consideration of the card layouts required and/or recommended for the PSL.

In general, it is good practice to adopt a standard ten column identification field with column nine reserved for a deck number and column ten reserved for card number. Columns one through eight would usually contain a unique subject identification number and other information upon which sorts of the data files are likely to be based. This information would be repeated on each card of a multiple-card file.

Also, it is most practical to punch variable fields without decimals because the Format Card (See Section 3 of this manual for a complete description of this card) permits the insertion of decimals in variables. Trailing and embedded zeros in variable fields (i.e., 101, 10, 1.05) should be punched as zeros and not be left blank. Leading zeros (i.e., 081) need not be punched, however. Although minus ("–") signs are permissible, no plus ("+") signs are permitted in input to PSL programs.

The user interested in a more detailed discussion of punch card processing and data card layout should see Chapter 3 in D. J. Veldman, *FORTRAN Programming for the Behavioral Sciences*, Holt, Rinehart & Winston, 1967.

CONTROL CARD DESCRIPTIONS

In describing the control cards used with library programs, it is convenient to differentiate between "system" control cards and "program" control cards. By restricting descriptions of system control cards to the introductory sections of the manual, needless repetition of material is eliminated and the impact of operating system modifications upon documentation is minimized. Those program control cards common to most PSL programs are described below.

PSL USER'S MANUAL

3. The Control Cards of the PSL

There are two kinds of control cards necessary for utilization of the PSL; system control cards and program control cards. A deck ready for processing on the computer will always contain the following configuration of cards:

     a. // JOB T card
     b. // XEQ ON
     c. Account card
     d. // XEQ card for PSL program
     e. Program control cards and data
     f. // XEQ OFF
     g. Blank card

Items (a), (b), (c), (d), (f), and (g) are considered system control cards and are described in detail below. Certain program control cards will also be described below; however, other program control cards are described in the write-up for the PSL program with which they are used.

SYSTEM CONTROL CARDS

a. // JOB T Card. This card causes output from processing of your deck to appear on a new page and tells the computer which disk cartridges will be used during your job. A list of PSL programs and the disk cartridges on which they are stored is posted in the computer room. The following // JOB T cards are used for PSL programs stored as indicated.

For a program stored on disk cartridge 0001:

1	2	3	4	5	6	7	8
/	/		J	O	B		T

For a program stored on disk cartridge 0002:

1	2	3	4	5	6	7	8	9	10	11	12	13	14	15	16	17	18	19
/	/			J	O	B		T			0	0	0	1		0	0	0	2

For a program stored on disk cartridge 0003:

1	2	3	4	5	6	7	8	9	10	11	12	13	14	15	16	17	18	19	20	21	22	23	24
/	/		J	O	B		T			0	0	0	1		0	0	0	2		0	0	0	3

b. // XEQ ON Card. This card, part of the Peabody Time Accounting System, is punched exactly as follows:

Column	Entry
1–2	//
3	Blank
4–6	XEQ
7	Blank
8–9	ON
10–80	Blank

c. Account Card. This card, also part of the Peabody Time Accounting System, is punched as follows:

Column	Entry
1	. (period)
2–6	Account Number
7–10	Blank
11–25	Name, problem title, etc. Information punched in these columns will be printed verbatim on the output and entered in the job log.
26–30	Blank
31–34	Code Word
35–80	Blank

For further information concerning the Peabody Time Accounting System and/or obtaining an account number, see the *Peabody Computer Center User's Manual* [Appendix B].

d. // XEQ Card for PSL Program. This card is punched as follows (Program C01 is used as an example):

Column	Entry
1–2	//
3	Blank
4–6	XEQ
7	Blank
8–10	Program Name (e.g., C01)
11–80	Blank

f. // XEQ OFF Card. This card, another part of the Peabody Time Accounting System, is punched exactly as follow:

Column	Entry
1–2	//
3	Blank
4–6	XEQ
7	Blank
8–10	OFF
11–80	Blank

PROGRAM CONTROL CARDS

Title Card. One or more title cards may be included in the job deck for most PSL programs. The function of the Title Card is to provide the PSL user a means of labeling his computer output so that he can easily identify the salient characteristics of the sample upon which the analysis was performed. Any convenient alphabetic and/or numeric information may be punched on the Title Card, and an unlimited number of Title Cards may be included in the job deck. *All of the Title Cards must be grouped together in the job deck, and an alphabetic (nonnumeric) character must be punched in column one of each Title Card.* A possible configuration of Title Cards might be

**ALL FEMALES FROM THE JOHNSON HIGH SCHOOL
SAMPLE COLLECTED ON 1/1/72**

It is easy to underestimate the importance of an adequate set of Title Cards: many a researcher has discovered to his dismay that the computer output that seemed clear in May is unusably obscure by August.

Parameter Card. This card provides numerical information required for the successful execution of a specific program. Unless otherwise noted, all entries are *right-justified* in four-column fields. This means, for example, that if columns five through eight are allocated for a parameter and the actual number to be punched is 37, the number should be punched so that the 7 is in column eight. Leading blanks (in this example, columns five and six) are treated as zeros.

Format Card. This card describes the layout (= "format") of input data for the computer. Specifications appearing on the Format Card identify the data card columns from which scores are to be read and the data card columns which are to be skipped.

X notation is used to indicate that certain columns of each input data card file are to be skipped. The form of X notation is symbolized as

$$nX$$

where n is a number greater than 0 and equal to the number of columns to be skipped. The following examples illustrate X notation:

Specification	Interpretation
6X	skip 6 columns
41X	skip 41 columns
1X	skip 1 column

F notation is used to describe variable (score; data) fields. The form of F notation is symbolized as

$$rFw.d$$

w = width of the variable field in columns.
d = number of digits after the decimal.
r = number of similar fields in succession.
Consider the following examples:

Specification	Interpretation
6F10.4	6 fields in succession, each 10 columns wide with 4 digits after the decimal.
4F2.0	4 fields in succession, each 2 columns wide with no decimal places.
F5.1	1 field five columns wide with one digit after the decimal.

The decimal point need not actually appear in the variable field. Instead, the d specification may be used to insert a decimal in variable fields where none is punched. The following table illustrates how the digits 73145 may be read, using F notation:

Specification	Resulting Number
F5.0	73145.0
F5.1	7314.5
F5.2	731.45
F5.3	73.145
F5.4	7.3145
F5.5	.73145

If a decimal point is actually punched in a variable field it will override the F notation field specification. For example, if 31.63 were read by F5.0, the number actually read would be 31.63.

Successive Format Card specifications are separated by commas, and parentheses are used to enclose the entire set of entries on the Format Card. Consider the following examples and their interpretation:

This Format Card tells the computer to skip 10 columns, read two adjacent fields three columns wide with a decimal point before the last digit, skip one column, and read one field six columns wide with two digits after the decimal point.

Spacing on the Format Card is optional, and the Format Card above is equivalent to

or

The Format Card

is equivalent to the Format Card

since one level of nesting of parentheses may be used. This strategy can greatly simplify the writing of complex format specifications. More than one set of nested parentheses may appear in a set of format specifications:

Multiple cards per subject may be handled two ways. If each card in the subject's file contains the same layout, format specifications describing the first card of the set will be automatically repeated by the computer. The other way to handle this situation is through the use of the slash ("/"). A slash appearing in a set of format specifications is interpreted as "skip to the next card." Accordingly, the Format Card

would be interpreted as follows:

Skip ten columns, read three one-column fields without decimals; skip to the next card, skip ten columns, read two two-column fields without decimals. The slash can also be used to skip one or more cards in each subject's card set without reading any information from them:

The above Format Card would cause the computer to skip the first and third cards of each subject's four-card set.

If the set of specifications needed to describe the input to a particular PSL program will not fit on a single Format Card, it is possible to continue the format specifications on a second Format Card. To do this, simply punch an asterisk (*) in column 80 of the first Format Card and continue the specifications on the second card. The last closing parenthesis would, of course, appear on the second card.

Variable Names Cards. These cards permit the specification of names or identification for variables. Variable names so specified will appear as labels on the program output. Any alphabetic and/or numeric label may be used, and each label should be left-justified in its field.

Twenty-column fields for each variable name are allocated as follows:

Columns	Entry
1–20	Name for variable 1
21–40	Name for variable 2
41–60	Name for variable 3
61–80	Name for variable 4

The specification of variable names may continue on additional cards as necessary.

Particular care is taken in describing the relatively complicated format card. A simplified explanation with extensive examples is presented in the hope of providing the user with enough understanding to successfully create a format card for his particular set of data. It is, however, the format card that mystifies users more than any other system or program control card, and staff support is often required during the user's first attempts to write format specifications.

A Program Using Variable Format

Program C01 is a PSL routine featuring the simulated variable format described in the preceding chapter. Sections of the write-up for program C01 are presented below.

GENERAL DESCRIPTION

Under this heading are included brief statements of the statistics computed, the program's output, salient parameter limitations, and run times for sample problems of varying size.

PSL USER'S MANUAL: CLASS C—CORRELATION

C01:   PEARSON PRODUCT-MOMENT CORRELATION

1. General Description

a. This program computes means, sigmas, covariances, and Pearson product-moment correlation coefficients. Punched output of the covariance matrix and/or the correlation matrix may be obtained.

b. The output consists of:
   (1) Means.
   (2) Sigmas.
   (3) Covariance matrix (optional).
   (4) Correlation matrix (optional).
   (5) The upper triangle (including the main diagonal of variances) of the covariance matrix punched as a continuous vector of $V(V + 1)/2$ elements (where V = number of variables), 7 elements per card (optional).
   (6) The upper triangle (including the main diagonal of ones) of the correlation matrix punched as a continuous vector of $V(V + 1)/2$ elements (where V = number of variables), 7 elements per card (optional).

c. Limitations:
   (1) The number of variables may not exceed 50.
   (2) The number of subjects may not exceed 9,999.

d. Estimation of running time per data set: The table below presents approximate run times in seconds for various combinations of number of variables and number of subjects. In each case, each subject's card file consisted of one card, and all optional printed output (but not punched output) was specified.

	Number of Variables		
Number of Subjects	5	20	50
20	72	180	684
100	108	216	972
1,000	288	972	4,176

ORDER OF CARDS IN JOB DECK

This section of the write-up provides a ready reference for the user as he assembles his job deck. The user is referred to the introductory section of the PSL manual for descriptions of those cards common to most PSL routines.

2. Order of Cards in Job Deck

    a. System control cards. (See Section 3* of this manual.)
    b. Title card(s). (Optional; see Section 3 of this manual.)
    c. Parameter card.
    d. Format card(s).
    e. Variable names card(s). (Optional; see Section 3 of this manual.)
    f. Data cards.
    g. Blank cards for punched output (if specified).
    h. Blank card (stops program).
    i. System control cards. (See Section 3 of this manual.)
    Items (b) through (g) may be repeated for additional sets of data.

CARD PREPARATION

In this section the fields of the parameter card are defined, the details of the format card are presented, and the general organization of the data cards is briefly reviewed. A formula for calculating the number of blank cards for punched output (if specified) is also given. The inclusion of the blank cards in the job deck is necessitated by the IBM 1442 card READ/PUNCH unit found on the typical IBM 1130. This device features a single input path passing through both a READ and a PUNCH station. Accordingly, blank cards for punched output must be inserted in the job deck.

3. Card Preparation

    c. Parameter Card

Columns	Entry
1–4	Number of variables
5–8	Number of subjects
9–12	Variable names option: 1 if variable names cards are to be read 0 if not

---

*Refers to the PSL manual section reproduced on pp. 40–45 of this text.

13-16	Output options for matrix of covariances:

0 = no output
1 = print only
2 = punch only
3 = print and punch

17-20	Output options for matrix of correlations: (same as above)
21-80	Blank

d. Format Card(s). The Format Card(s) must specify only variable fields in *F*-notation. Use *X*-notation to skip each subject's identification field. See Section 3 of this manual for a detailed discussion of the Format Card(s).

f. Data Cards. The same number of variables must appear in each subject's card file, and each variable must appear in the same location in each subject's card file.

g. Blank Cards for Punched Output (if specified). If punched output is specified, $V(V + 1)/14$ (where V = number of variables) blank cards for each matrix specified must follow the data cards.

REFERENCES

This section is included for the user desiring a review of the statistics computed by the program. It is highly desirable to provide specific page or chapter references in documents readily available to most users.

4. References

The Pearson product-moment correlation coefficient is discussed in W. L. Hays, *Statistics for Psychologists*, New York: Holt, Rinehart and Winston, 1963, pp. 492-538; and the concept of covariance is presented in Paul Horst, *Psychological Measurement and Prediction*, Belmont, Calif.: Wadsworth, 1966, p. 88.

COMPUTATIONAL NOTES

If confusion concerning the exact formula implied by a statistical term could arise, the actual formulas (or their equivalents) used should be included under computational notes.

5. Computational Notes

The mean is defined as $\Sigma X_i/N$ ($i = 1,N$), where X_i is the score for the ith subject and N = number of subjects.

The sigma is defined as the square root of $\Sigma X_i^2/N - M^2$ $(i = 1,N)$, where X_i and N are defined as above, and $M =$ mean.

The covariance between X and Y is defined as $\Sigma(X_i - M_x)$ $(Y_i - M_y)/N$, $(i = 1,N)$, where X_i and N are defined as above, $M_x =$ mean of X, and $M_y =$ mean of Y.

All output has been rounded.

This program will accept input from disk. Experienced programmers desiring to use this feature may obtain a special write-up from the Computer Center.

Any other computational peculiarities of the program should also be described in this section. The note concerning the possibility of disk input to the program alerts the experienced FORTRAN programmer to a program feature judged too complex for use by the typical researcher.

```
*** SYSTEM CONTROL CARDS ***
TEST PROBLEM -- 1/1/72
    5   10    1    3    3
(5X, 5F1.0)
ITEM A              ITEM B            ITEM C            ITEM D
ITEM E
01    45123
02    54321
03    12345
04    47823
05    54321
06    98765
07    12345
08    65123
09    45321
10    78531
(SIX BLANK CARDS FCR PUNCHED CUTPUT)
(BLANK CARD TO STCP PRCGRAM)
*** SYSTEM CONTROL CARDS ***
```

Figure 3-1. Sample Problem for Program C01

SAMPLE PROBLEM AND OUTPUT

The sample problem (Figure 3-1) is designed to exercise most of the features of the program without generating so much output that the size of the write-up becomes excessive. As mentioned earlier, the sample problem and output (Figure 3-2) provide the user an example to study as he learns how to set up the program. In addition, the actual form of the output listed in the general description of the program is available for inspection. Finally, the sample problem provides a convenient means of testing the program.

SPECIAL WRITE-UP FOR DISK INPUT

Intended for experienced FORTRAN programmers, this supplementary write-up presents the essential features of disk input to program C01. A general description of the provision for disk input is given, a modified card order is shown, and brief description of the technique for interfacing a permanent file with program C01 is given. The sample program with comment cards is a vital part of the write-up.

```
PROGRAM CC1, PEARSON PRODUCT-MOMENT CORRELATION, VERSION OF 1/1/72
GEORGE PEABODY COLLEGE COMPUTER CENTER, NASHVILLE, TENNESSEE

TEST PROBLEM -- 1/1/72

PARAMETER CARD ENTRIES
COLUMNS  ENTRY  MEANING

  1-4      5    NUMBER OF VARIABLES
  5-8     10    NUMBER OF SUBJECTS
  9-12     1    VARIABLE NAMES OPTION
 13-16     3    COVARIANCE MATRIX OUTPUT OPTIONS...
                  0 = NO OUTPUT
                  1 = PRINT ONLY
                  2 = PUNCH ONLY
                  3 = PRINT AND PUNCH
 17-20     3    CORRELATION MATRIX OUTPUT OPTIONS...
                  (SAME AS ABOVE)

FORMAT CARD(S) =

(5X, 5F1.0)


VARIABLE           MEAN           SIGMA          DESCRIPTION

   1              4.600          2.332           ITEM A
   2              5.000          2.049           ITEM B
   3              3.700          2.193           ITEM C
   4              2.900          1.300           ITEM D
   5              2.800          1.661           ITEM E


COVARIANCES

                              1         2         3         4         5
   1 ITEM A                5.440     4.000     1.780     0.560    -1.080
   2 ITEM B                4.000     4.200     2.800     0.300    -0.800
   3 ITEM C                1.780     2.800     4.810     1.170     0.540
   4 ITEM D                0.560     0.300     1.170     1.690     1.580
   5 ITEM E               -1.080    -0.800     0.540     1.580     2.760


CORRELATIONS

                              1         2         3         4         5
   1 ITEM A                1.000     0.837     0.348     0.185    -0.279
   2 ITEM B                0.837     1.000     0.623     0.113    -0.235
   3 ITEM C                0.348     0.623     1.000     0.410     0.148
   4 ITEM D                0.185     0.113     0.410     1.000     0.732
   5 ITEM E               -0.279    -0.235     0.148     0.732     1.000



                   *** PUNCHED OUTPUT ***

C MAT   1   5.4400     4.0000     1.7800     0.5600    -1.0800     4.2000
C MAT   2   0.3000    -0.8000     4.8100     1.1700     0.5400     1.6900
C MAT   3   2.7600
R MAT   1   1.0000     0.8368     0.3480     0.1847    -0.2787     1.0000
R MAT   2   0.1126    -0.2350     1.0000     0.4104     0.1482     1.0000
R MAT   3   1.0000
```

Figure 3-2. Sample Problem Output for Program C01

PSL USER'S MANUAL: CLASS C—CORRELATION

USING PROGRAM C01 WITH DISK FILES

1. General Description

a. Program C01 may be run with data input from disk rather than cards. It is the user's responsibility to create and load his own permanent disk files and to deposit in a standard temporary file data for input to Program C01.

b. The output is essentially the same as that available with data input from cards.

c. Limitations with disk input:

(1) The number of variables may not exceed 50.

(2) The number of subjects may not exceed 1,000.

d. Estimation of running time per disk file: The table below presents approximate run times in seconds for various combinations of number of variables and number of subjects. In each case, the required number of variables and subjects were selected from the first N (where N is the number of subjects represented in a row of the table) records of a 1 000-record permanent file containing 160-word records. All optional printed output was specified; also, run times include compilation of the program to select subjects for the analysis.

		Number of Variables		
		5	20	50
	20	108	180	720
Number of Subjects	100	108	252	972
	1,000	360	972	4,176

2. Order of Cards in Job Deck

a. System control cards and user program to place data in standard temporary file.

b. Title card(s). (Optional; see Section 3 of this manual.)

c. Parameter card.

d. Variable names card(s). (Optional; see Section 3 of this manual.)

e. Blank cards for punched output (if specified).

f. Blank card (stops program).

g. System control cards (see Section 3 which follows).

Each disk file must be processed in a separate run.

3. Card Preparation

System Control Cards and User Program to Place Data in Standard Temporary File. The user is assumed to be familiar with IBM 1130 FORTRAN and the IBM 1130 Disk Monitor System, including the creation and management of permanent disk files.

The user-written program must contain a temporary file defined by the statement

DEFINE FILE 1 (1000,100,U,N1)

File 1 must be the first temporary file created in the program.

The user's program must also contain the statement

COMMON NS

and be compiled in standard precision with one-word integers. At the time the user's program transfers control to Program C01 with the statement CALL LINK (C01), the variable NS must contain the number of subjects whose data are stored in real mode in the first NS records of file 1.

The cards below could constitute item (a) in a job deck. The user desires to run an analysis on 25 variables for all the males represented in the permanent file SUBJ on cartridge 0333.

```
// JOB T  0001 0002 0333
// XEQ ON
 (account card)
// FOR
*ONE WORD INTEGERS
*LIST SOURCE PROGRAM
*IOCS  (DISK)
       DIMENSION X(25), IT(100)
       COMMON NS
       DEFINE FILE 1 (1000,100,U,N1), 2 (300,100,U,N2)
C
C      INITIALIZE NS.
C
       NS = 0
C
C      TRANSFER SCORES FOR MALES TO FILE 1.
C
       DO 20 I = 1,300
       READ (2'I) (IT(J), J = 1,100)
C
C      TEST FOR MALE (IT(4) = 1).
C
       IF (IT(4) - 1) 20,5,20
C
C      MALE FOUND - INCREMENT NS AND COPY SCORES INTO
       FILE 1.
```

```
C
    5   NS = NS + 1
        DO 10 J = 1,25
   10   X(J) = IT(J+13)
   15   WRITE (1'NS) (X(J), J = 1,25)
C
C       MALE NOT FOUND - CONTINUE LOOP.
C
   20   CONTINUE
C
C       TRANSFER CONTROL TO PROGRAM C01.
C
        CALL LINK (C01)
        END
// XEQ          1
*FILES(2,SUBJ,0333)
```

c. Parameter Card. Only those columns whose entries differ in the case of disk input are shown.

Column	Entry
5–8	Blank
21–24	1 (signifies disk input)

Since our concern in this chapter is focused on the form of documentation rather than on programming techniques, no discussion of the method for using data from disk as input to a library program appears here. Instead, the reader is referred to Chapter 5.

A Program Using Variable Location Cards

Program D03 performs a relatively simple task for which the variable location card is well adapted.

GENERAL DESCRIPTION

As described in this section, program D03 is written to accept nonnumerical codes without protest; furthermore, the program distinguishes blanks from zeros. Both features would be impossible to implement using simulated variable format (subroutines FMAT and TAKE).

PSL USER'S MANUAL: CLASS D—DESCRIPTIVE STATISTICS

D03: FREQUENCY TABULATION FOR
SINGLE-COLUMN VARIABLES

1. General Description

a. This program accepts data cards containing single-character alpha-numeric (letters of the alphabet, numerals and special characters) codes for a series of variables and compiles, for each variable, frequency of occurrence of blank, zero through nine and "other" (all nonblank, nonnumerical codes). These frequencies of occurrence are converted to percentages of total number of subjects and printed along with means and sigmas based only on the codes one through nine.

b. The output consists of:

(1) For each variable, frequencies of occurrence of blank, zero, one, two, three, four, five, six, seven, eight, nine, and "other" (all nonblank, nonnumerical codes).

(2) For each variable, percentages of the total number of subjects whose codes fall in each of the above categories.

(3) For each variable, a mean and sigma based only on nonzero numerical codes.

c. Limitations:

(1) The number of variables may not exceed 100.

(2) The number of subjects may not exceed 9,999.

(3) The number of cards per subject may not exceed 5.

(4) Each variable may consist only of single-character codes.

d. Estimation of running time per data set: The table below presents approximate run times in seconds for various combinations of number of variables and number of subjects. In each case, each subject's card file consisted of two cards.

		Number of Variables		
		20	80	100
Number of Subjects	50	72	154	180
	200	154	252	252
	1,000	468	576	648

ORDER OF CARDS IN JOB DECK

The similarity of this section for program D03 to the same section for program C01 is an indication of the degree to which the features and concepts learned in mastering one library program transfer to other programs in the same library. The only new program control card is the variable location card.

2. Order of Cards in Job Deck

    a. System control cards. (See Section 3 of this manual.)
    b. Title card(s). (Optional; see Section 3 of this manual.)
    c. Parameter card.
    d. Variable location card(s).
    e. Variable names card(s). (Optional; see Section 3 of this manual.)
    f. Data cards.
    g. Blank card (stops program).
    h. System control cards. (See Section 3 of this manual.)

Items (b) through (f) may be repeated for additional sets of data.

CARD PREPARATION

The main item of interest in this section is the description of the variable location card. The simplicity of the explanation is striking when compared to the lengthy introductory section on variable format.

3. Card Preparation

    c. Parameter Card

Columns	Entry
1–4	Number of variables
5–8	Number of subjects
9–12	Variable names option:
	1 if variable names cards are to be read
	0 if not
13–16	Number of cards per subject
17–80	Blank

    d. Variable Location Card(s). Each of these cards corresponds to one of the data cards in a subject's card set. In other words, if the data cards include three cards per subject, three Variable Location Cards are required.

There is a one-for-one correspondence between Variable Location Card columns and data card columns. A number 1 (one) is punched in each Variable Location Card column which corresponds to a data card column in which a variable's codes are punched. Other Variable Location Card columns are left blank. There must, therefore, be V ones punched in the Variable Location Card columns (where V = number of variables).

If each subject's card set includes one or more cards from which no codes are to be read for a given computer run, there must nevertheless exist a Variable Location Card for each card of the subject card set. In this case, one or more Variable Location Cards would be completely blank.

f. Data Cards. The same number of variables must appear in each subject's card file, and each variable must appear in the same location in each subject's card file.

REFERENCE AND COMPUTATIONAL NOTES

These sections parallel very closely the corresponding sections in the write-up for program C01.

4. Reference

Frequency distributions, the mean, and the sigma are discussed in Janet T. Spence, B. J. Underwood, C. P. Duncan, and J. W. Cotton, *Elementary Statistics*, New York: Appleton-Century-Crofts, 1968, Chapters 2, 5, and 6.

5. Computational Notes

Only those subjects with non-zero numerical codes (= "scores") are included in the computation of the mean and sigma for each variable.

The mean is defined as $\Sigma X_i/N$ ($i = 1,N$), where X_i is the score for the ith subject and N = number of subjects.

The sigma is defined as the square root of $\Sigma X_i^2/N - M^2$ ($i = 1,N$), where X_i and N are defined as above, and M = mean.

All output has been rounded.

SAMPLE PROBLEM AND OUTPUT

Again an attempt is made to present a problem displaying most of the characteristics of the program. See Figures 3-3 and 3-4.

```
*** SYSTEM CONTROL CARDS ***
TEST DATA, 1/1/72
    5   10    1    1
      11111
QUESTIONNAIRE ITEM 1SEX (1 = M, 2 = F) CLASSIFICATION      POLITICAL PARTY
CANDIDATE PREFERRED
01   11231
02   124L2
03   11211
04   11111
05   41132
06   12121
07   32432
08   425
09   00123
10   21322
(BLANK CARD)
*** SYSTEM CONTROL CARDS ***
```

Figure 3-3. Sample Problem for Program D03

```
PROGRAM D03, FREQUENCY TABULATION FOR SINGLE-COLUMN VARIABLES, VERSION OF 1/1/72
GEORGE PEABODY COLLEGE COMPUTER CENTER, NASHVILLE, TENNESSEE

TEST DATA, 1/1/72

PARAMETER CARD ENTRIES
COLUMNS  ENTRY  MEANING

  1-4      5    NUMBER OF VARIABLES
  5-8     10    NUMBER OF SUBJECTS
  9-12     1    VARIABLE NAMES OPTION
 13-16     1    NUMBER OF CARDS PER SUBJECT

TABLE OF FREQUENCIES (ROWS = VARIABLES)
```

	BLANK	0	1	2	3	4	5	6	7	8	9	OTHER	DESCRIPTION (IF ANY)
1	0	1	5	1	1	2	0	0	0	0	0	0	QUESTIONNAIRE ITEM 1
2	0	1	5	4	0	0	0	0	0	0	0	0	SEX (1 = M, 2 = F)
3	0	0	4	2	1	2	0	0	0	0	0	0	CLASSIFICATION
4	1	0	2	3	3	0	0	0	0	0	0	1	POLITICAL PARTY
5	1	0	4	4	1	0	0	0	0	0	0	0	CANDIDATE PREFERRED

```
TABLES OF PERCENTAGES
```

VARIABLE	BLANK	0	1	2	3	4	DESCRIPTION (IF ANY)
1	0.00	10.00	50.00	10.00	10.00	20.00	QUESTIONNAIRE ITEM 1
2	0.00	10.00	50.00	40.00	0.00	0.00	SEX (1 = M, 2 = F)
3	0.00	0.00	40.00	20.00	10.00	20.00	CLASSIFICATION
4	10.00	0.00	20.00	30.00	30.00	0.00	POLITICAL PARTY
5	10.00	0.00	40.00	40.00	10.00	0.00	CANDIDATE PREFERRED

VARIABLE	5	6	7	8	9	OTHER	DESCRIPTION (IF ANY)
1	0.00	0.00	0.00	0.00	0.00	0.00	QUESTIONNAIRE ITEM 1
2	0.00	0.00	0.00	0.00	0.00	0.00	SEX (1 = M, 2 = F)
3	10.00	0.00	0.00	0.00	0.00	0.00	CLASSIFICATION
4	0.00	0.00	0.00	0.00	0.00	10.00	POLITICAL PARTY
5	0.00	0.00	0.00	0.00	0.00	0.00	CANDIDATE PREFERRED

```
MEAN, SIGMA, AND LOCATION OF EACH VARIABLE IN SUBJECT CARD SET
(BLANK, 0, AND OTHER EXCLUDED FROM COMPUTATION OF STATISTICS)
```

VARIABLE	CARD	COLUMN	MEAN	SIGMA	DESCRIPTION (IF ANY)
1	1	5	2.000	1.247	QUESTIONNAIRE ITEM 1
2	1	6	1.444	0.497	SEX (1 = M, 2 = F)
3	1	7	2.400	1.428	CLASSIFICATION
4	1	8	2.125	0.781	POLITICAL PARTY
5	1	9	1.667	0.667	CANDIDATE PREFERRED

Figure 3-4. Sample Problem Output for Program D03

SUMMARY

The goal of the author of documentation should be minimization of the need for staff consultation in support of a particular library program and the maximization of library user independence.

The sample documentation presented in this chapter illustrates one effort in pursuit of this goal. Probably the easiest way to begin production of documentation is to work from a model; it is hoped that this chapter provides such a model.

4 Programming the Small Computer

It is not uncommon for the FORTRAN programmer first confronting a medium- to small-scale computing system such as the IBM 1130, Digital Equipment Corporation PDP-11, or Hewlett-Packard 2100 series to throw up his hands in despair at the thought of ever running a nontrivial program for statistical analysis. Even relatively experienced programmers—particularly if they learned FORTRAN in the context of a large-scale computing system—consistently underestimate and underutilize the capabilities of these potentially powerful systems. Often the programs developed for use with such machines reflect a sense of confinement and Spartanism growing out of the programmer's acute awareness of the apparent limitations of the hardware. Given the fact that the operating systems supplied with the "minicomputers" of the 1970s are as sophisticated as the operating systems supplied with the most powerful computers available only 10 or 15 years earlier, the production of severely circumscribed applications software betrays the limitations of programmer rather than inadequacies of hardware.

The techniques presented in this chapter are intended to aid the FORTRAN programmer in circumventing some of the difficulties encountered in implementing statistical procedures on a computing system with limited core memory. Although there is a sizable repertory of programming techniques available to the programmer using even an unadorned computing system without magnetic tape or disk, much of the material presented in this chapter is intended for application in an operating environment in which mass storage is available.

59

PROBLEM SEGMENTATION

One approach to the management of programs too large to fit in available core memory is the identification of problem segments that may be divided into separate modules for processing. This may be accomplished in various ways, depending upon the nature of the operating system and the type of mass storage available.

Use of Intermediate Output

It is sometimes possible to use intermediate output for communication among separate programs. For example, a set of programs to perform principal components analysis with analytic rotations might be organized according to the scheme appearing in Figure 4-1.

The first program in the set would compute an intercorrelation matrix. The output from this routine would then serve as input to the next program, a routine to compute principal components. The transfer of information between programs would proceed in a similar fashion through the remaining stages of analysis. The medium for intermediate output could consist of punched paper tape, punched cards, or—in the case of somewhat more sophisticated computing systems—magnetic tape. The program to compute principal components would be written to read the correlation matrix as data. The use of magnetic tape for storage of intermediate output would permit stacking of the series of programs for batch processing.

If, on the other hand, punched paper tape or cards were used for intermediate output, the user of the programs would be required to retrieve the intermediate output for each stage of the analysis and insert it in the appropriate location in the job deck for the next stage of analysis.

Use of Mass Storage and Program Segmentation

Most computer systems featuring magnetic disk or tape storage afford the programmer the capability to set up linkages among program segments so that continuous processing can occur, with control of the system passing automatically between segments. We shall make a distinction between segmentation involving only mainline-to-mainline chaining (where the terms *mainline* and *main program* are synonymous) and segmentation involving subprograms within a single mainline.

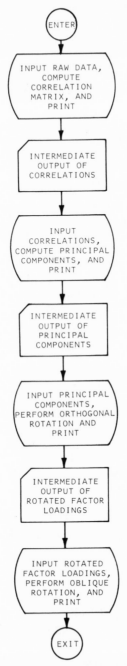

Figure 4-1. System of Programs to Perform Principal Components
Analysis with Analytic Rotations

MAINLINE-TO-MAINLINE CHAINING

This type of segmentation may be thought of as resembling the use of intermediate output in that the results of processing in one mainline are transferred to another mainline; however, in this case, COMMON, rather than intermediate output, is used to pass information from one mainline to the next. The COMMON area in core memory is saved when one mainline transfers control to another, so data residing in variables in COMMON are available to successive mainlines; however, each successive mainline can redefine the size of the COMMON area. For example, if program ONE contained the statement

COMMON KARDS, LINES, X(100), Y(100)

and program TWO contained the statement

COMMON KARDS, LINES, X(100)

any data in the array Y would not be saved for program TWO. Because the variables appearing in corresponding positions in COMMON need not be the same from mainline to mainline, the COMMON area may be regarded as a dynamic repository of parameters and data providing the capability for communication among the modules of a system of mainline programs.

Consider the following set of programs written for use on an IBM 1130 disk operating system:

```
C  MAINLINE ONE
   COMMON NV, NS, X(100), SUMX(100), KF(180)
   (Input of control cards and tests for validity
    of parameters)
   CALL LINK (TWO)
   END

C  MAINLINE TWO
   COMMON NV, NS, X(100), SUMX(100), KF(180)
   (Input of data and accumulation of ΣX for each
    variable)
   CALL LINK (THREE)
   END

C  MAINLINE THREE
   COMMON NV, NS, XBAR(100), SUMX(100)
   (Computation and printing of means)
   CALL LINK (ONE)
   END
```

These programs would be stored on disk under the names ONE, TWO, and THREE, respectively. Transfer of control between routines is accomplished by execution of the CALL LINK statement. Execution of CALL LINK causes the main program whose name appears within the parentheses to be fetched from disk storage. Loading and/or execution then commences (depending upon whether the program named has been stored on disk by means of a *STORE or *STORECI control record; see Chapter 5). Although the CALL LINK statement is specific to the IBM 1130, comparable facility exists for most other disk-operating systems.

The dynamic nature of COMMON is illustrated by the different configuration of variables and arrays in each of the mainlines. Mainline ONE handles input of control cards. Since values for NV (number of variables) and NS (number of subjects) are needed in all three mainlines, NV and NS appear first in COMMON. The array KF (storage for variable format specifications) is needed in both mainline ONE and mainline TWO, so it appears in COMMON in both routines; however, after input of data is accomplished in mainline TWO, KF is no longer needed. Accordingly, it has been omitted from COMMON in mainline THREE, thus making more core memory available for other variables, arrays, and object code. It is possible, of course, to include arrays in a DIMENSION statement in each of the mainlines. Such an array would not be available outside the mainline in which it was dimensioned.

The capability for associating different variable or array names with the same area in COMMON has been exploited in the case of mainline TWO and mainline THREE. In mainline TWO the array X appears; in mainline THREE the same area is occupied by XBAR. This change in name affords the programmer a name with improved mnemonic value without increased cost in core memory. Because machines like the IBM 1130 use (in standard precision) twice as much core memory for real variables and arrays as for integer variables and arrays, it is also possible to rename an area in COMMON as follows:

| (Mainline A) | COMMON X(100) |
| (Mainline B) | COMMON KX(200) |

A word of caution is in order here, however. In order to avoid problems arising from the way in which core memory is allocated to integer and real arrays in COMMON during program loading, it is good practice to situate in the latter part of the COMMON statement any real arrays for which integer arrays are later to be substituted. For example, a reasonable configuration of COMMON statements might be

COMMON KARDS, LINES, NV, NS, X(100), Y(100)

and

COMMON KARDS, LINES, NV, NS, X(100), KY(200)

On the other hand, the statements

COMMON KARDS, LINES, NV, NS, X(100), Y(100)

and

COMMON KARDS, LINES, NV, NS, KX(200), Y(100)

could lead to improper alignment of the array Y between the two programs containing the COMMON statements.

The present description of mainline-to-mainline chaining has focused on the IBM 1130. This feature is implemented somewhat differently on other computers, and the documentation concerning FORTRAN and the operating system for a given computer should be consulted for detailed instructions. It should be noted that mainline-to-mainline chaining is practical even in the case of tape operating systems, since the time overhead associated with tape manipulation, while an order of magnitude greater than that associated with disk, is substantially less than the processing delays occasioned by actually inputting an object deck for each successive mainline in processing.

CAUSING SUBPROGRAMS TO SHARE CORE MEMORY

The FORTRAN programmer working with an IBM 1130 disk operating system can avail himself of still another means of reducing the core memory required for the storage of object code. Consider the following highly simplified program:

```
C  PROGRAM SHARE
   DIMENSION X(1000), Y(50)
C  DEFINE I/O DEVICE NUMBERS.
   KARDS = 2
   LINES = 5
C  INPUT A CONTROL CARD TO DEFINE NV AND NS.
   CALL CONTR (KARDS,NV,NS)
C  INPUT DATA.
   CALL DATA (KARDS,NV,NS,X)
C  PERFORM CALCULATIONS.
   CALL CALCU (NV,NS,X,Y)
C  PRINT RESULTS.
   CALL RESUL (LINES,NV,Y)
   CALL EXIT
   END
```

Because the subroutines CONTR, DATA, CALCU, and RESUL do not call one another, it is possible to cause them to share a single area in core memory through the use of the IBM 1130 disk operating system LOCAL facility. The use of LOCAL (an acronym standing for "load-on-call") causes selected subprograms (subroutines and/or functions) to reside on disk until referenced by the mainline. At that time, the subprogram referenced is loaded from disk into a special area of core memory, overlaying any LOCALed subprogram previously occupying that area. Linkage between the mainline and LOCALed subprograms is handled automatically by the system, and the LOCAL facility can be implemented in most cases by the addition of only a single control card (see Chapter 5).

The savings in core memory requirements achieved through the use of LOCAL can be enormous. For example, suppose the subroutines called by the preceding program SHARE required core memory (in IBM 1130 words) as follows:

Subroutine	Core
CONTR	270
DATA	350
CALCU	500
RESUL	450

Since the area of core memory shared by the subprograms must be as large as the core requirements of the largest subprogram, approximately 500 words of core would be required to set up the LOCAL area (more is actually required because special system routines for the management of the LOCALed subprograms must be included in the core load). The 1070 words of core memory required for CONTR, DATA, and RESUL would be saved through the use of LOCAL. Once the LOCAL area has been established, additional LOCALed subprograms may be referenced by the mainline program at little additional expense with respect to core memory. Through the use of LOCAL the FORTRAN programmer can compile and execute programs totaling many thousands of words of object code—and on a system with only 8192 words of core memory!

Although one LOCALed FORTRAN subprogram cannot reference another, there is no restriction upon the number of times a LOCALed subprogram may be referenced. On the other hand, the time overhead introduced by the disk activity associated with all LOCAL loads requires that the programmer exercise some care in organizing his programs. Suppose, for example, a programmer has divided a computational sequence into three subprograms, CAUL1, CAUL2, and CAUL3, referenced as follows:

.

.

.

```
         CALL CAUL1 (A,B,C)
         DO 100 I = 1,1000
         X(I) = CAUL2 (A,B,I)
   100   Y(I) = CAUL3 (X(I),B)
```

.

.

.

While the disk activity associated with the single loading of CAUL1 would go unnoticed, the repeated swapping of CAUL2 and CAUL3 for each other would consume a relatively substantial amount of time. An alternative organization of program modules to avoid reference to more than one LOCALed subprogram within a loop would be preferable. For similar reasons, the inclusion of disk READ and WRITE statements and references to LOCALed subprograms within a loop should be avoided.

USE OF COMMON VERSUS ARGUMENT LISTS

Where possible, communication among subprograms and the mainline should be accomplished using COMMON rather than argument lists. In addition to affording the opportunity to redefine areas in COMMON within a given program, subprograms communicating via COMMON require less core storage than subprograms featuring an argument list. An "ideal" mainline might appear as follows:

```
   C   PROGRAM MAIN
       COMMON KARDS, LINES, NV, NS, X(1000)
       CALL SUB1
       CALL SUB2
       CALL SUB3
       CALL SUB4
       CALL SUB5
       CALL EXIT
       END
```

Each subroutine called by program MAIN would contain a COMMON statement. Because SUB1, SUB2, SUB3, SUB4, and SUB5 do not call one another, it would be possible to conserve core memory through the LOCAL facility; furthermore, debugging would be expedited by the division of program code into more manageable modules. If the use of LOCAL is not sufficient to alleviate core memory shortages, the use of COMMON will have facilitated division of program MAIN into two or more LINKed mainlines.

Separating Phases of Processing for Maximum Use of Core

The allocation of phases of processing among mainlines ONE, TWO, and THREE in the discussion of mainline-to-mainline chaining arises from considerations extending beyond mere organizational convenience. An awareness of some of these considerations can help the FORTRAN programmer in his attempts to optimize his routines for maximum use of core memory.

SOFTWARE IMPLICATIONS OF FORTRAN STATEMENTS*

In order to minimize hardware costs, the designers of small computing systems have relegated to software many of the operations ordinarily handled by hardware on larger systems. For example, most small computing systems lack floating-point (real arithmetic) hardware; the FORTRAN statement

$$X = A / B$$

causes the FORTRAN compiler to generate a reference to a subprogram for floating-point division in the object code it produces in processing source statements. Integer arithmetic, on the other hand, is performed by hardware on many small computing systems (e.g., Digital Equipment Corporation systems with extended arithmetic hardware element; Hewlett-Packard systems with extended arithmetic instructions; IBM 1130).

The READ and WRITE statements in FORTRAN are also implemented by system software. In general, each device referenced requires the inclusion of system routines in the core load when the main program is executed. Accordingly, the object code generated by a program both reading cards and printing on a line printer will require more core memory than the object code generated by a program in which only the reading of cards occurs.

ORGANIZING PROGRAMS TO MINIMIZE CORE-RESIDENT SOFTWARE

Taking into account the software implications discussed briefly above, it can be seen that input, output, and computation could be separated into separate mainlines to avoid the simultaneous existence in core memory of system

---

*General discussions of FORTRAN optimizations may be found in F. Gruenberger, *Computing: A Second Course*, San Francisco: Canfield Press, 1971, chap. 6; and C. Larson, "The Efficient Use of FORTRAN," *Datamation*, vol. 17 (1971), 24–31.

software required to support the FORTRAN statements associated with these phases of processing. In fact, a reexamination of mainlines ONE, TWO, and THREE discloses that optimal separation of phases of processing has not been accomplished.

Mainline ONE involves both input of control cards and tests for validity of parameters (with associated echoing of parameters); therefore, all three phases of processing are included in mainline ONE. There are, however, advantages to placing control card input, echoing, and tests in a single, separate mainline. First, on systems lacking the LOCAL facility, the FORTRAN subprograms (e.g., subroutine START, as described in Chapter 2) required to support the responsibilities apportioned to mainline ONE would appear only in its core load. Second, it is often possible to avoid in a program like mainline ONE the inclusion of large arrays required during subsequent data input and computation, since the arrays needed during input of control cards are needed only for storage of control card specifications. Third, although some computation is involved in the testing of the validity of parameters supplied on control cards, only integer arithmetic need be used. This is convenient because, as mentioned earlier, many small computing systems perform integer arithmetic with hardware rather than software.

Although it is not apparent from a first glance at mainline TWO, all three phases of processing could be present. If tests for various input errors (e.g., invalid character in numeric input field) have been included in the input sequence, then output statements for printing of error messages must also be included. It is true that only a limited number of the system subprograms required for floating-point arithmetic would appear in the core load, since the accumulation of ΣX would probably involve a statement of the form

$$\mathrm{SUMX(J) = SUMX(J) + X(J)}$$

This statement would generate a reference to a system subprogram for floating-point addition, but system subprograms for such floating-point operations as subtraction, division, or multiplication would not be included in the core load. Our example programs are restricted in scope for clarity, but it is, of course, quite likely that most statistical routines would require the full range of floating-point operations.

In mainline THREE, input has been avoided entirely, but both floating-point arithmetic and output are required.

While the organization of mainlines ONE, TWO, and THREE is a step in the right direction, it is clear that phases of processing would be better separated. Figure 4-2 portrays an alternative organization of the functions of mainlines ONE, TWO, and THREE. In this case, phases of processing have been distributed among five mainlines.

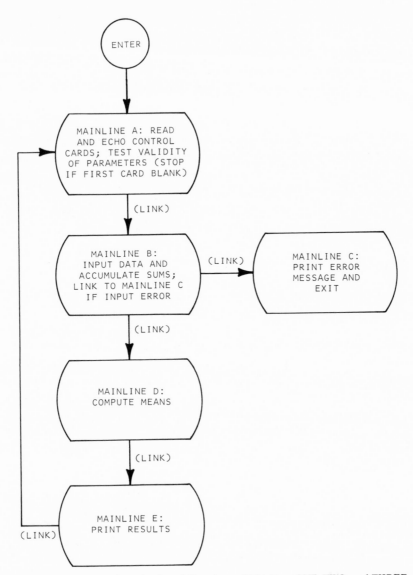

Figure 4-2. Alternative Organization of Functions of Mainlines ONE, TWO, and THREE

Mainline A corresponds to mainline ONE in the previous organizational scheme. No attempt to separate phases of processing has been made in the case of mainline A because more core memory is available in mainline A than in later stages of processing. It would, of course, be possible to divide mainline A into several mainlines if core memory restrictions dictated.

Printing has been eliminated from mainline B. Because the detection of an error during input is to result in program termination, it is feasible to branch to mainline C for printing of an error message and termination.

The previous mainline THREE has also been subdivided. Computation is handled in mainline D, and printing occurs in mainline E. After execution of mainline E, control is returned to mainline A, which reads an additional card and decides whether to terminate execution (in the case of a blank card) or begin processing a new set of data.

The program organization appearing in Figure 4-2 represents a rather extreme extension of a strategy for conserving core memory requirements. In many cases it will be possible to implement statistical procedures without resorting to such drastic measures; nevertheless, the separation of phases of processing can be a highly useful trick of the trade in the repertory of the programmer of a small computer.

CREATING MULTIPLE ENTRY POINTS IN A MAINLINE

Although mainline-to-mainline chaining ordinarily results in execution of each successive mainline beginning with the first statement in that mainline, the need for an organization resembling calls to subroutines sometimes arises. For example, suppose in a limited-core situation that several matrices computed during processing are to be printed and that a mainline has been written to perform that function. Suppose further (for the sake of illustration) that each matrix is computed in the same mainline. What is needed then is a means of chaining to the matrix-printing mainline and returning to the appropriate point in the original matrix-computing mainline. The statements COMMON and a computed GO TO provide a solution:

```
C       MAINLINE TO BEGIN PROCESSING
        COMMON KARDS, LINES, NV, NS, NN
        Define KARDS and LINES, device numbers for card reader
        and line printer.
        Input control cards defining NV, number of variables, and
        NS, number of subjects.
        NN = 1
        CALL LINK (MATS)
        END

C       PROGRAM MATS
        COMMON KARDS, LINES, NV, NS, NN, X(1000)
        GO TO (5,50,100,200,300), NN
    5   Input data and compute first matrix.
        NN = 2
```

```
          CALL LINK (PRINT)
    50    Compute second matrix.
          NN = 3
          CALL LINK (PRINT)
   100    Compute third matrix.
          NN = 4
          CALL LINK (PRINT)
   200    Compute fourth matrix.
          NN = 5
          CALL LINK (PRINT)
   300    CALL EXIT
          END

C         PROGRAM PRINT
          COMMON KARDS, LINES, NV, NS, NN, X(1000)
          Print matrix.
          CALL LINK (MATS)
          END
```

The variable NN in COMMON is used to store a pointer indicating where processing is to begin upon entry into program MATS. The mainline that begins processing sets NN = 1. Accordingly, the first transfer of control to MATS results in a branch to statement 5 for data input and computation of the first matrix. Then NN is reset to 2 and control is transferred to program PRINT. Program PRINT prints the matrix and transfers control back to program MATS. Since NN is now 2, execution of the computed GO TO results in a branch to statement 50. In a similar fashion, computation and printing of matrices continues until statement 300, a CALL EXIT, is encountered. If provision for stacking sets of data for analysis had been desired, statement 300 could have been made a CALL LINK transferring control back to the first mainline for input of additional control cards.

The effect of this organization is to cause program PRINT to function like a LOCALed subroutine; however, system software associated with printing would appear in the program PRINT core load rather than in the core load for program MATS. Accordingly, this scheme could prove useful in situations where conservation of core memory is of extreme importance or when computer systems lack the LOCAL facility.

As in previous discussions of mainline-to-mainline chaining, the details of syntax (i.e., CALL LINK) apply to the IBM 1130, and equivalent procedures for other systems must be substituted in programs written for use on those systems.

STORING MATRICES EFFICIENTLY

The matrices required for many statistical techniques present a special problem to the programmer of the small computer. The author of a program that must include, for example, an array to store NV variables for each of NS subjects is likely to spend a considerable amount of time agonizing over the optimal dimensions for a two-dimensional array X. If X is to contain NS rows and NV columns, what are the "best" values of NS and NV to use in the DIMENSION statement when the product of NS and NV cannot exceed, say, 1500?

Compressed Storage of Nonsymmetric Matrices

Consider the two arrays diagrammed in Figure 4-3. The two-dimensional array X contains scores on three variables for each of five subjects. If we were writing a program to perform a statistical analysis requiring the data to reside in core memory, we might decide that limits of 100 and 15 for the number of subjects and the number of variables, respectively, represent an optimal compromise within the limitation of 1500 for the total number of elements in X. In this case, X would be dimensioned 100 by 15.

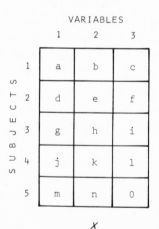

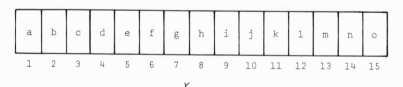

Figure 4-3. Storage of Equivalent Arrays

While the data in Figure 4-3 would certainly fit comfortably within these limitations, such data sets as 200 subjects by 5 variables or 75 subjects by 20 variables would exceed program limitations, even though neither data set contained more than 1500 data points.

A solution to this problem is suggested by the arrangement of the elements of the one-dimensional array Y in Figure 4-3. Instead of storing the 15 data points in the two-dimensional array X, successive rows of X are stored in Y as indicated (where the letters in X correspond to the letters in Y). This packing of data points in the lower-order elements of Y eliminates the arbitrary limitations on the number of subjects and the number of variables imposed by the use of a two-dimensional array.

READING A NONSYMMETRIC MATRIX INTO COMPRESSED STORAGE

The sequence of statements

```
      NSTAR = 1
      DO 100 I = 1,NS
      NSTOP = NSTAR + NV - 1
      READ (KARDS,KF) (Y(J), J = NSTAR,NSTOP)
  100 NSTAR = NSTAR + NV
```

would read NV scores for each of NS subjects into the array Y (where the dimension of Y was at least NS * NV). The variable KARDS is assumed to have been defined as the device number for the card reader, and the one-dimensional array KF is assumed to contain variable format specifications.

If the simulation of variable format described in Chapter 2 is in use, the sequence of statements

```
      NE = 0
      DO 100 I = 1,NS
      CALL TAKE (KF,DATA,NV,KARDS,LINES)
      DO 100 J = 1,NV
      NE = NE + 1
  100 Y(NE) = DATA(J)
```

would accomplish the same task as the statements given for "real" variable format. In this case, KARDS, NV, NS, and Y are defined as in the earlier input sequence. The variable LINES is assumed to have been defined as the device number for the line printer, and KF is assumed to contain the format specifications encoded by an earlier call to subroutine FMAT. The one-dimensional array DATA is used for temporary storage of each subject's scores, so the number of elements dimensioned for DATA would determine the number of variables the program could handle (assuming DATA to be dimensioned with fewer elements than Y).

REFERENCING ELEMENTS OF A COMPRESSED NONSYMMETRIC MATRIX

Referring again to arrays X and Y in Figure 4-3, Y(NE) corresponds to X(I,J), where NE = (I - 1) * 3 + J. In general, the expression for NE is

$$NE = (I - 1) * NV + J$$

where NV has been defined as the number of variables. If, for example, we wanted to store the row and column sums of our data matrix (stored in the one-dimensional array Y) in the arrays RSUM and CSUM, respectively, the statements

```
      DO 100 J = 1,NV
      CSUM(J) = 0.0
      DO 100 I = 1,NS
      NE = (I -1) * NV + J
100   CSUM(J) = CSUM(J) + Y(NE)
      DO 105 I = 1,NS
      RSUM(I) = 0.0
      NE = (I -1) * NV
      DO 105 J = 1,NV
      NE = NE + J
105   RSUM(I) = RSUM(I) + Y(NE)
```

would do the job. It would, of course, be necessary to dimension CSUM and RSUM with at least NV and NS elements, respectively. The separation of the calculation of NE into two arithmetic statements in the DO 105 loops is designed to minimize the number of times the expression

$$(I - 1) * NV$$

would be evaluated.

PRINTING A COMPRESSED NONSYMMETRIC MATRIX

Because the input data matrix is seldom printed, no sequences of code for printing a compressed nonsymmetric matrix are presented here. It should, however, be noted that subroutine MPRT could be easily modified to incorporate the expression for NE given above. The argument list of subroutine MPRT would be modified so that the calling sequence would be

CALL MPRT (X,NR,NC,LINES)

with the arguments defined as in the discussion of MPRT; and the third statement after statement number 55 in subroutine MPRT would be changed from

$$KK = L + ND * (M - 1)$$

to

$$KK = (L - 1) * NC + M.$$

Compressed Storage of Symmetric Matrices

When the matrix that must be stored in core memory is a symmetric matrix such as a correlation matrix, the scheme suggested in Figure 4-4 may be used. As portrayed by the lower-case letters in R, $r_{ji} = r_{ij}$. Accordingly, the arrangement of elements from R in the one-dimensional array RR retains all information required to reconstruct R.

Figure 4-4. Vector Representation of a Symmetric Matrix

READING A SYMMETRIC MATRIX INTO COMPRESSED STORAGE

If variable format is to be used, the statements

$$NTOT = (NV * NV - NV) / 2 + NV$$
$$READ (KARDS,KF) (RR(I), I = 1,NTOT)$$

would cause the computer to read the upper triangle from a symmetric matrix, where the input cards had been punched in a layout equivalent to that generated by the statements

WRITE (KPNCH,50) ((R(I,J), J = I,NV), I = 1,NV)
50 FORMAT (10X, 7F10.4)

(where KPNCH has been defined as the logical unit number for the card punch). In actual practice, some sort of identification code and sequence number would be punched in columns 1 through 10 of each card containing matrix elements. The variable format specifications stored in the array KF referenced in the READ statement above would, of course, be

(10X, 7F10.4)

regardless of whether or not columns 1 through 10 of each input card contained a sequence number.

An equivalent input sequence using simulated variable format would be

NTOT = (NV * NV - NV) / 2 + NV
CALL TAKE (KF,RR,NTOT,KARDS,LINES)

(assuming that the array KF had been defined by a call to subroutine FMAT).

REFERENCING ELEMENTS OF A COMPRESSED SYMMETRIC MATRIX

It is probably more likely that a symmetric matrix would be defined by computation rather than input. Since, for example, one formula for the computation of the Pearson product-moment correlation coefficient[*] requires, for each pairing of variables X_i and X_j, ΣX_i, ΣX_j, ΣX_i^2, ΣX_j^2, and $\Sigma X_i X_j$, the following statements might be used to accumulate the required sums for all possible pairings of NV variables for NS subjects:

```
      DO 100 I = 1,NS
      NE = 0
      READ (KARDS,KF) (X(J), J = 1,NV)
      DO 100 J = 1,NV
      SUMX(J) = SUMX(J) + X(J)
      DO 100 K = J ,NV
      NE = NE + 1
  100 RR(NE) = RR(NE) + X(J) * X(K)
```

The elements of the arrays SUMX and RR are assumed to have been previously initialized to zero. As in the case of previous examples, the READ statement could be replaced by a call to subroutine TAKE if simulated variable format was to be used. After execution of this sequence, the ith element of SUMX will contain ΣX for the ith variable; the elements of RR will correspond to a symmetric matrix of sums of squares and sums of products as diagrammed in Figure 4-5. Since $\Sigma X_i X_i = \Sigma X_i^2$, the main diagonal of R contains sums of squares

---

[*]Guilford, J. P., *Fundamental Statistics in Psychology and Education*. New York: McGraw-Hill Book Company, 1965, Chapter 6.

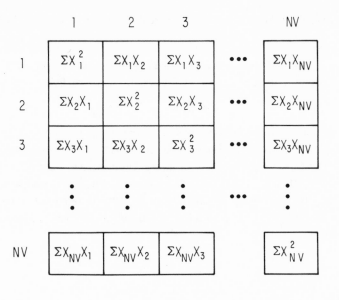

R

Figure 4-5. A Matrix of ΣX_i^2 and $\Sigma X_i X_j$

for the NV variables. Because $\Sigma X_i X_j = \Sigma X_j X_i$, the matrix R is symmetric about the main diagonal. Actually, as we shall see in Chapter 8, a slightly different computational scheme is used in calculating $r_{x_i x_j}$. The sequence of statements presented above involves a method simplified for the sake of clarity of presentation.

If nonsequential access to the elements of RR corresponding to certain elements of R is required, successive increments of a counter such as NE are inappropriate. Instead, a variable NTOT is computed by the statement

$$\text{NTOT} = (\text{NV} * \text{NV} - \text{NV}) / 2 + \text{NV}$$

and the value of NE for RR(NE) corresponding to R(I,J) is defined by one of two statements, depending upon the relationship between I and J:

```
        IF (I - J) 50,50,55
    50  II = I
        JJ = J
        GO TO 60
    55  II = J
        JJ = I
    60  NE = NTOT - NV + JJ - (NV - II) * (NV - II + 1) / 2
```

Of course RR would be dimensioned with at least NTOT elements.

PRINTING A COMPRESSED SYMMETRIC MATRIX: USE OF SUBROUTINE DPRT

This routine may be used to print a compressed symmetric matrix without variable names. The calling sequence for subroutine DPRT is

CALL DPRT (X,N,LINES)

where X is the one-dimensional real array that contains an N by N symmetric matrix in compressed form. The variable LINES is the device number for the line printer. The matrix represented in the array X is printed in eight-column blocks.

The printing of a compressed symmetric matrix with associated variable names may be accomplished using subroutine DPRTN described in Appendix A. Listings of both subroutine DPRT and subroutine DPRTN also appear in Appendix A.

Variable Dimensioning

Many FORTRAN compilers permit the use of nonsubscripted integer variable names in place of integer constants in the DIMENSION statements of subprograms. Any integer variable name used in this fashion must appear in the argument list of the subprogram. The use of the feature permits the FORTRAN programmer to create subprograms more general than routines in which the dimensions of matrices are fixed. Figure 4-6 presents Version A of a simple subroutine for the addition of two matrices. Because the amount of memory allocated for the arrays X, Y, and Z is actually controlled by the calling program,

```
C
C       SUBROUTINE TO ADD TWO MATRICES, VERSION A
C
        SUBROUTINE ADD (X,Y,Z,NR,NC,ND)
C
C       X AND Y = TWO MATRICES TO BE ADDED.
C       Z = MATRIX TO CONTAIN SUM.
C       NR = NUMBER OF ROWS IN X, Y, AND Z.
C       NC = NUMBER OF COLUMNS IN X, Y, AND Z.
C       ND = NUMBER OF ROWS DIMENSIONED FOR X, Y, AND Z.
C
        DIMENSION X(ND,1), Y(ND,1), Z(ND,1)
        DO 5 I = 1,NR
        DO 5 J = 1,NC
      5 Z(I,J) = X(I,J) + Y(I,J)
        RETURN
        END
```

Figure 4-6. Subroutine to Add Two Matrices, Version A

the DIMENSION statement in the subprogram is necessary only to provide information concerning the number of subscripts to be used in referencing X, Y, and Z within the subprogram and to establish, through the use of ND in the DIMENSION statement, correspondence between array elements in the subprogram and array elements in the calling program. Although X, Y, and Z must have the same number of rows, there is no restriction on the number of columns in the three matrices. If separate integer variable names (NDX, NDY, and NDZ, for example) in the argument list had been used in a DIMENSION statement of the form

DIMENSION X(NDX,1), Y(NDY,1), Z(NDZ,1)

there would have been no need for X, Y, and Z to have been dimensioned with the same number of rows in the calling program.

Unfortunately, the FORTRAN available on some small computers such as the IBM 1130 does not include variable dimensioning as described above. In these more restricted versions of FORTRAN there is, however, a way to simulate variable dimensioning without actually using an integer variable name in the DIMENSION statement of the subprogram. As mentioned earlier, the DIMENSION statement of a subprogram does not, in the case of arrays appearing in the argument list, actually reserve core memory; furthermore, although an array may have been dimensioned with two subscripts in the calling program, the same array may be equated to a subprogram dummy argument referred to as a one-dimensional array within the subprogram. Version B of our simple subroutine to add two matrices makes use of both characteristics of FORTRAN. As can be seen in Figure 4-7, X, Y, and Z are each dimensioned with one element within

```
C
C       SUBROUTINE TC ADD TWC MATRICES, VERSION B
C
        SUBROUTINE ADD (X,Y,Z,NR,NC,ND)
C
C       X AND Y = TWC MATRICES TC BE ADDED.
C       Z = MATRIX TO CONTAIN SUM.
C       NR = NUMBER OF ROWS IN X, Y, AND Z.
C       NC = NUMBER CF CCLUMNS IN X, Y, AND Z.
C       ND = NUMBER CF RCWS DIMENSIONED FOR X, Y, AND Z.
C
        DIMENSION X(1), Y(1), Z(1)
        CO 5 I = 1,NR
        CO 5 J = 1,NC
        IJ = I + ND * (J - 1)
      5 Z(IJ) = X(IJ) + Y(IJ)
        RETURN
        END
```

Figure 4-7. Subroutine to Add Two Matrices, Version B

Version B. The statement just before statement 5 computes the single subscript, which will permit reference to the appropriate elements of X, Y, and Z. To understand why the statement before statement 5 works, we must consider how two-dimensional arrays appear in core memory. When we equate the calling program two-dimensional array X with a one-dimensional dummy argument, we achieve the same effect as the statement

EQUIVALENCE (X(1,1), Y(1)).

Suppose the array X is dimensioned 5 by 3 in the calling program and the set of data for a particular execution of the program has resulted in definition of three rows and two columns of X as portrayed in Figure 4-8. Because FORTRAN allocates core storage for two-dimensional arrays in column order, elements from X appear in Y as shown. The FORTRAN statement

$$IJ = I + ND * (J - 1)$$

formalizes this relationship and provides a value for IJ such that Y(IJ) corresponds to X(I,J) (in Figure 4-8, ND = 5).

	1	2	3
1	a (3.)	f (2.)	k
2	b (1.)	g (7.)	l
3	c (10.)	h (3.)	m
4	d	i	n
5	e	j	o

X

a 3.	b 1.	c 10.	d	e	f 2.	g 7.	h 3.	i	j	k	l	m	n	o
1	2	3	4	5	6	7	8	9	10	11	12	13	14	15

Y

Figure 4-8. Equivalent Elements in One- and Two-Dimensional Arrays

SUMMARY

Techniques for achieving better utilization of the limited core memory of the small computer have been the central focus of this chapter.

One very useful strategy in implementing relatively large programs on small computer systems is some form of problem segmentation—use of intermediate output, for example, or segmentation of the program itself, as in the case of mainline-to-mainline chaining.

Where matrices are involved in the analysis, it is often possible to store them more efficiently through representation in one-dimensional arrays. Specific examples of one-dimensional representation of matrices were presented along with subroutines to facilitate printing.

5 The IBM 1130 Disk

Although most of the techniques presented in this book are applicable to computer systems of widely varying size, we have made a particular effort to emphasize the feasibility of implementing reasonably ambitious statistical procedures on even small-scale computers. It is true, however, that some sort of rapid-access mass storage is of enormous help (as demonstrated in Chapter 4). In particular, a disk operating system converts the small computer into a very comfortable operating environment.

In Chapter 4 we considered strategies applicable to a number of small computing systems. In this chapter we study in detail a few IBM 1130 disk features of particular relevance to the FORTRAN programmer engaged in the creation of a statistical program library. It is not intended here to provide comprehensive coverage of disk-related techniques for the IBM 1130. Instead, only features and techniques judged to be particularly useful in support of the types of routines discussed in this book have been included.

USE OF DISK FOR TEMPORARY FILES

It is not unusual for a particular statistical technique to require the storage of data in quantities exceeding the capacity of core memory available on the IBM 1130. In such an event the disk can be used to provide comparatively large amounts of random-access mass storage.

Organization of Disk Storage

The IBM 2315 disk cartridge provides 512,000 16-bit words of interchangeable mass storage divided into 1600 320-word sectors. Disk files may contain only uniform records less than or equal to a sector in length. Our initial consideration of disk files will be limited to files existing only during execution of the program in which they are defined.

Establishing and Referencing a Temporary Disk File

In order to use a temporary disk file within our program, we must provide the FORTRAN compiler with information concerning the internal organization of the file.

THE DEFINE FILE STATEMENT

The general form of this statement is

$$\text{DEFINE FILE } n \ (r, w, U, v)$$

where n = integer constant $(1 \leqslant n \leqslant 32,767)$ denoting a symbolic file number to be referenced in subsequent disk I/O statements

r = integer constant $(1 \leqslant r \leqslant 32,767)$ specifying the number of records in the file

w = integer constant $(1 \leqslant w \leqslant 320)$ specifying the number of words per record

v = nonsubscripted integer variable

The letter U, specifying unformatted disk I/O, must always appear in the third position within the parentheses and has been included in the syntax of the statement to maintain "upward compatibility" with IBM 360 FORTRAN. The integer variable v is known as the *associated variable* and will be discussed in detail later. We shall also have more to say later about w, the number of words per record.

As an example, consider the statement

$$\text{DEFINE FILE } 99 \ (125, 80, U, N99)$$

which establishes file 99 with 125 records of 80 words each. Two or more files may be defined in a single statement:

$$\text{DEFINE FILE } 1 \ (10, 320, U, N1), \ 8 \ (1000, 100, U, N8)$$

File 1 would contain 10 320-word records, and file 8 would contain a thousand 100-word records. The pattern of names for the associated variable (file 99, N99; file 1, N1; etc.) is merely one possible convention; any valid nonsubscripted integer variable name would do.

The statement DEFINE FILE appears only in a mainline program immediately following any specification statements:

1. Type statements
2. External statements
3. Dimension statements
4. Common statements
5. Equivalence statements
6. Data statements
7. Define file statements

THE DISK WRITE STATEMENT

The statement used to transfer information from core memory to the disk has the general form

$$\text{WRITE } (n'r) \text{ } list$$

where n = integer constant or variable specifying a file number previously declared in a DEFINE FILE statement

r = integer constant, variable, or expression specifying the number of the record at which transmission of information is to begin

$list$ = list of variables and array elements whose current values are to be transferred

The statement

$$\text{WRITE } (101'I) \text{ } (X(J), J = 1, NV)$$

would cause the current values of elements one through NV of the array X to be transmitted to file 101, beginning with record I. The disk WRITE is nondestructive with respect to core memory in the same fashion as an ordinary FORTRAN write: The statement simply causes a *copy* of information in core memory to be created elsewhere (on disk, in this case). On the other hand, the disk WRITE *does* replace any existing information on disk with new information. The following are also examples of valid disk WRITE statements:

$$\text{WRITE } (NFILE'1) \text{ } A, B, KK$$
$$\text{WRITE } (99'J+5) \text{ } I, (X(M), M = 1, 5)$$

THE DISK READ STATEMENT

Information may be transferred from disk to core memory by a statement of the form

READ ($n'r$) *list*

where n, r, and *list* are defined as for the disk WRITE statement. A copy of the information stored on disk is simply created in variables and/or arrays in core memory.

THE ASSOCIATED VARIABLE

After execution of each disk READ or WRITE, the nonsubscripted integer variable v is automatically updated to point to the next available record in the file. For example, suppose file 99 had been established by the statement

DEFINE FILE 99 (100,20,U,N99)

and the statement

55   WRITE (99'6) (KK(I), I = 1,35)

appeared in the same program. Assuming one-word integers, statement 55 would cause the current values of the 35 elements of KK to be stored in records 6 and 7 of file 99 (with five words of record 7 unused). After execution of statement 55, the value of the associated variable N99 would be 8. Then N99 could be used as r in a disk READ or WRITE. The associated variable, then, provides a convenient means of keeping track of where I/O operations should next begin in sequential access to a disk file in situations where I/O lists may cause information to spill across records. If the associated variable is to be used in subprograms, it must be transmitted via COMMON. Use of the associated variable will be illustrated in various applications in subsequent sections of this chapter.

Storing Data as One Record per Subject

A common library program requirement is the storage of information for each subject in a sample. This could be accomplished by the following sequence of statements (where NS and NV are the number of subjects and the number of variables, respectively):

DO 5 I = 1,NS
CALL TAKE (KF,X,NV,KARDS,LINES)
5   WRITE (999'I) (X(J), J = 1,NV)

In each cycle of the DO loop, subroutine TAKE inputs NV scores according to the format specifications stored in the array KF. The NV elements of X are stored in successive records of file 999 by statement 5.

DETERMINING DEFINE FILE PARAMETERS

File 999 must, of course, be established by a DEFINE FILE statement. The appropriate entry for r, the number of records in the file, is fairly obvious. If the plan is to store one subject's scores in each record, then r must be as large as the largest anticipated value of NS, the number of subjects. But what about w, the number of words per record? If our program has been compiled in standard precision, each real array element occupies two words of core memory. Accordingly, the NV elements of X occupy 2 * NV words. Our entry for w must therefore be twice the largest anticipated value of NV. If our program is designed to handle data sets no larger than 1000 subjects and 50 variables, a suitable DEFINE FILE statement would be

DEFINE FILE 999 (1000,100,U,N999)

In the case of subject-by-subject storage of information, the appropriate value of w is generally determined by examination of the I/O lists associated with the disk READ and WRITE statements referencing the file. In general, the I/O lists of the disk READ and WRITE statements will be identical, and either statement may be considered. Suppose the statement

WRITE (3'I) (X(J), J = 1,NV), TOT, NN

is to be used to reference a file for NS subjects. Assuming that NS $\leqslant$ 500, NV $\leqslant$ 100, standard precision, and that both TOT and NN are nonsubscripted variables, we can make up Table 5-1. The total of 203 words for each of our 500 or fewer subjects could be stored in the file created by the following statement:

DEFINE FILE 3 (500,203,U,N3)

BLOCKING SEVERAL SUBJECTS' DATA FOR MORE EFFICIENT CARD-TO-DISK TRANSFER

If the number of words of information to be stored for each subject is equal to, or slightly less than, a number dividing evenly into 320, it is possible to reduce the number of disk WRITE statements (and hence the run time) required during input of data from cards. Consider an example.

Table 5-1. Maximum Core Memory Required by a Sample I/O List

LIST ITEM	MODE	WORDS*
X(NV)	Real	200†
TOT	Real	2
NN	Integer	1
Total		203

*Using 16-bit IBM 1130 words and assuming two-word real variables or array elements and one-word integers.

†Assuming NV $\leqslant$ 100.

Up to 75 scores for each of up to 1200 subjects are to be stored on disk. We might first anticipate using the statement

$$\text{WRITE } (101\text{'I}) (X(J), J = 1,NV)$$

in a DO loop indexed from 1 to NS. For NV = 75 (and standard precision), there would be 150 words of information for each subject, leading us to define file 101 as follows:

$$\text{DEFINE FILE } 101(1200,150,U,N101)$$

We notice, however, that 150 is reasonably close to 160, one-half of a 320-word sector. Carrying our plan a step further, we note that a file of 1200 160-word records occupies as much disk storage as the following file:

$$\text{DEFINE FILE } 101 \ (600,320,U,N101)$$

Using an array BLOCK dimensioned with 960 elements and an array X dimensioned with 75 elements, we write the input sequence in Figure 5-1. Except for perhaps the last few subjects (if NS does not divide evenly by 12), we collect data from 12 subjects in BLOCK before each execution of statement 25.

After the data have been stored in file 101, our program can use CALL LINK to transfer control to another mainline in which the statement

$$\text{DEFINE FILE } 101 \ (1200,160,U,N101)$$

reconfigures the file so that access to a given subject's data is facilitated. Data for one subject can simply be read into a one-dimensional real array dimensioned 75, thus ignoring at least the last ten words in each record. See the later section, *Passing Temporary Files between Mainlines*, for additional comments related to this procedure.

```
C
C        BLOCKING INPUT DATA FOR STORAGE ON DISK
C
C        INITIALIZE ASSOCIATED VARIABLE, SUBJECTS COUNTER, AND
C           BLOCK LOCATION POINTER.
C
         N101 = 1
         N = 0
       5 LB = 0
C
C        INPUT A SUBJECT'S DATA.
C
      10 CALL TAKE (KF,X,NV,KARDS,LINES)
C
C        TRANSFER DATA TO BLOCK.
C
         DO 15 I = 1,NV
         NE = LB + I
      15 BLOCK(NE) = X(I)
C
C        INCREMENT N AND CHECK FOR LAST SUBJECT.
C
         N = N + 1
         IF (N - NS) 20,25,25
C
C        INCREMENT BLOCK LOCATION POINTER AND CHECK FOR END OF BLOCK.
C
      20 LB = LB + 80
         IF (LB - 960) 10,25,25
C
C        STORE BLOCK ON DISK.
C
      25 WRITE (101'N101) (BLOCK(I), I = 1,960)
C
C        RE-INITIALIZE BLOCK LOCATION POINTER IF MORE DATA TO COME.
C
         IF (N - NS) 5,30,30
      30 (NEXT STATEMENT IN PROGRAM)
```

Figure 5-1. Blocking Input Data for Storage on Disk

HANDLING I/O LISTS THAT EXCEED 320 WORDS PER SUBJECT

In the event that a relatively large amount of information must be stored for each subject, a little ingenuity is required. For example, the statement

$$\text{WRITE } (88'I) (X(J), J = 1, NV)$$

presents a problem if the maximum value of NV is greater than 160, since (in standard precision) 2 * NV words of information would be transmitted, and the value of w in a DEFINE FILE statement may not exceed 320.

As a first approximation to a solution, let us consider the management of a file for NV $\leqslant$ 200 and NS $\leqslant$ 500. We could set up a file by the statement

$$\text{DEFINE FILE 88 } (1000, 320, U, N88)$$

and transfer data from card to disk by the following sequences of statements:

```
        N88 = 1
        DO 5 I = 1,NS
        CALL TAKE (KF,X,NV,KARDS,LINES)
      5 WRITE (88'N88) (X(J), J = 1,NV)
```

The associated variable N88 is automatically set to the number of the 320-word sector at which transfer of the next NV scores is to begin. If NV is less than or equal to 160, each subject's scores would occupy 320 or fewer words of storage, and N88 would simply take on the values 1, 2, 3, . . . , NS. If, on the other hand, NV is greater than 160, the NV scores would spill across records, and N88 would assume the values 1, 3, 5, . . . , NS * 2 + 1. It is important to note that N88 must be initialized to 1 so that it is defined for the first cycle through the DO 5 loop. Random access to the set of NV scores for the Kth subject could be obtained through the following statements:

```
      IF (NV - 160) 75,75,80
  75  NREC = K
      GO TO 85
  80  NREC = (K - 1) * 2 + 1
  85  READ (88'NREC) (X(I), I = 1,NV)
```

While our first approximation would do the job, it is very wasteful of disk storage. We have defined file 88 as consisting of a thousand 320-word records, or a total of 320,000 words of disk storage, while our maximum of 200 scores for each of a maximum of 500 subjects amounts to a total of only 200,000 (200 scores * 2 words per variable * 500 subjects) words. With 200 scores per subject, 240 words of every other disk sector are unused.

A dramatic improvement in disk-storage efficiency can be achieved by setting w equal to the largest common divisor of 320 (the number of words in a sector) and 400 (the maximum number of words required for each subject's data). The appropriate value for w in this case is 80. Dividing 80 into 400, we find that each subject's data would require a maximum of five records. Accordingly, our revised file would have a total of 5 * NS records, or 2500 for NS = 500. The statement

DEFINE FILE 88 (2500,80,U,N88)

would set up the file, and the input sequence would be the same as for a record length of 320 words. The sequence of statements for random access to the set of NV scores for the Kth subject is, however, different:

```
       NN = NV * 2 / 80
       IF (NV * 2 - 80 * NN) 105,105,100
  100  NN = NN + 1
  105  NREC = NN * (K - 1) + 1
       READ (88'NREC) (X(I), I = 1,NV)
```

The first three statements determine the number of 80-word records occupied by the NV scores; statement 105 computes the appropriate record number.

If the largest common divisor of 320 and the maximum number of words of information to be stored for each subject is a relatively small number, the savings in disk storage may be offset by the increase in run time occasioned by the large number of disk transfers. For example, a maximum of 324 words of information to be stored for each of up to 500 subjects might seem to dictate a record length of 4 words, but this value of w is clearly unworkable. First of all, each subject's data would occupy 81 records, and a total of 40,500 records would be required—an invalid entry in a DEFINE FILE statement. Even if the maximum value for NS were within a workable range, records of 4 words each would be undesirable from a standpoint of run time. It would be preferable to use a compromise record length of 40. This record length would accommodate the maximum of 324 words per subject in 9 records with 36 unused words of disk storage per subject. Data for 500 subjects could then be stored in a file of 4500 records, or 180,000 total words. If NWT is the total number of words to be stored for one subject, NWR is the number of words per record, and NFILE is the number of the file in which information is stored, the sequence of statements for random access to the information for the Kth subject may be generalized as follows:

$$NN = NWT / NWR$$
$$IF\ (NWT - NWR * NN)\ 105,105,100$$
$$100\quad NN = NN + 1$$
$$105\quad NREC = NN * (K - 1) + 1$$
$$READ\ (NFILE\,'NREC)\ list$$

Temporary Storage of Arrays

Although an IBM 1130 with 8192 words of core memory cannot accommodate a FORTRAN program with two 1500-element real arrays in core memory at once, the need for the simultaneous existence of two large arrays sometimes arises. On such occasions the disk can easily be used as scratch storage.

FILE STRUCTURE

Using a standard record length of 320 words in order to maximize speed of transfer, the number of records is determined by dividing the number of words of storage required for the array by 320 and rounding up if there is a nonzero remainder. Table 5-2 presents some sample arrays, their memory requirements, and the number of 320-word records required for each. The apparent waste of disk storage is unavoidable; the IBM 1130 allocates space for disk files in sectors.

Table 5-2. Core and Disk Storage Requirements for Sample Arrays

ARRAY AND DIMENSIONS	MODE	CORE*	DISK RECORDS†
X(200)	Real	400	2
A(40,40)	Real	3200	10
KA(100,20)	Integer	2000	7
Q(10,10,20)	Real	4000	13
RR(50)	Real	100	1

*Core memory requirements in 16-bit IBM 1130 words, assuming two-word real-array elements and one-word integers.

†Disk storage requirements in 320-word records.

STORING AND RETRIEVING AN ARRAY

Figure 5-2 contains key statements from a program in which two 1500-element real arrays coexist. It is not, of course, necessary to dimension both arrays and subsequently equivalence them; however, this practice may be desirable if considerable mnemonic significance is attached to the two separate array names. Because the 1500 elements of X require 3000 words of storage, file 1 is assigned ten 320-word records. After the elements of X have been computed, the disk WRITE statement saves X on disk. The designation of record 1

```
C
C     EXAMPLE OF USE OF DISK FOR SCRATCH STORAGE
C
      DIMENSION X(1500), Y(1500)
      EQUIVALENCE (X(1),Y(1))
      DEFINE FILE 1 (10,320,U,N1)
      .
      .
      .
      (STATEMENTS TO COMPUTE ELEMENTS OF X)
      .
      .
      .
C
C     SAVE X ON DISK.
C
      WRITE (1'1) (X(I), I = 1,1500)
      .
      .
      .
      (STATEMENTS TO COMPUTE AND USE ELEMENTS
      OF Y)
      .
      .
      .
C
C     RECOVER X FROM DISK.
C
      READ (1'1) (X(I), I = 1,1500)
      .
      .
      .
      (REST OF PROGRAM)
```

Figure 5-2. Example of Use of Disk for Scratch Storage

in the WRITE statement merely loads X into file 1, beginning with record 1; the current values of the elements of X spill across all ten records of file 1 automatically. With X safely stored on disk, the same area in core memory is reused as the array Y. Finally, X is recovered from disk and processing continues.

STORAGE OF VARIABLE NAMES

The contents of the variable names cards described in Chapter 2 can conveniently be stored on disk. A disk file with one 20-word record per variable is defined (our example assumes a maximum of 100 variables):

$$\text{DEFINE FILE 3 (100,20,U,N2)}$$

Transfer of the variable names for NV variables from card to disk is accomplished by the following statements:

```
      DO 50 I = 1,NV,4
      READ (KARDS,45) (KK(J), J = 1,80)
   45 FORMAT (80A1)
   50 WRITE (2'I) (KK(J), J = 1,80)
```

Since a complete card is read and its contents transferred to disk with each cycle through the loop, it is important to allocate enough records in the DEFINE FILE statement. If the maximum number of variables is not an even multiple of 4, the smallest even multiple of 4 greater than the maximum number of variables must be entered as the number of records in the DEFINE FILE statement. Otherwise, an attempt to reference a nonexistent record could occur.

Subroutine DPRTN and MPRTN print matrices with associated variable names. Both routines read variable names from disk and are described in Appendix A.

The FIND Statement

Considerable savings in disk access time may be realized through use of the FIND statement in certain situations.

SYNTAX

The general form of the statement is

$$\text{FIND } (n'r)$$

where n is an integer constant or variable referring to a file, and r is an integer, constant, variable, or expression specifying the number of a record in file n.

The FIND statement causes the read/write heads of the disk to seek record r of file n while processing of subsequent program statements continues simultaneously. If a FIND statement can be executed enough program statements ahead of a disk READ or WRITE referencing the specified record, a fraction of a second can be saved. If the use of the FIND statement occurs in a loop, the time savings can be significant.

EXAMPLE OF USE

Suppose the numbers of records from disk file 99 selected for printing have been stored in the first NSEL elements of the one-dimensional integer array KK. Since successive elements of KK could contain such widely separated record numbers as 1, 21, 103, 516, 713, etc., considerable disk access time could be spent in moving between records. Assuming that file 99 has been established by the statement

DEFINE FILE 99 (1000,20,U,N99)

information could be read from disk and printed by the following statements:

```
      NREC = KK(1)
      NN = NSEL - 1
      DO 5 I = 1,NN
      READ (99'NREC) (KDATA(J), J = 1,20)
      NREC = KK(I+1)
      FIND (99'NREC)
   5  WRITE (LINES,10) (KDATA(J), J = 1,20)
  10  FORMAT (1X,20I6)
      READ (99'NREC) (KDATA(J), J = 1,20)
      WRITE (LINES,10) (KDATA(J), J = 1,20)
```

While each relatively slow WRITE to the line printer is in progress, the disk READ/WRITE heads are being positioned over the next record specified in KK. The READ and WRITE statements following the loop are necessary to print the record specified in KK(NSEL).

FIND AND PROGRAM SEGMENTATION

The FIND statement should not be used when any disk operation (involving the disk containing the file referenced in the FIND statement) occurs between execution of the FIND and the subsequent related disk READ or WRITE. For example, on a one-disk IBM 1130, the FIND statement should not be used if a call to a LOCALed subprogram (see Chapter 4 and the discussion below) occurs between it and the next disk READ or WRITE.

Use of Disk as an Extension of Core Memory

So far we have considered the use of the disk for temporary storage of input data or of arrays previously defined in core memory. Now we shall conceptualize a disk file as a disk-resident array and examine methods of manipulating that array without making it entirely core-resident.

DISK STORAGE OF A COMPRESSED SYMMETRIC MATRIX

As noted in Chapter 4, the total number of elements, NTOT, in a one-dimensional representation of a symmetric matrix in compressed form may be computed by the formula

$$NTOT = (NV * NV - NV) / 2 + NV$$

where NV is the dimension of the symmetric matrix. Since each element of a real array in standard precision occupies two words of memory, we could represent the matrix in a disk file with NTOT records of two words each. For example, a disk file to hold a symmetric matrix of maximum dimension 200 could be defined by the statement

DEFINE FILE 7 (20100,2,U,N7)

and NREC, the number of the record corresponding to element I,J of the symmetric matrix, could be computed by a slightly modified version of a sequence of statements given in Chapter 4:

```
        IF (I - J) 50,50,55
   50   II = I
        JJ = J
        GO TO 60
   55   II = J
        JJ = I
   60   NREC = NTOT - NV + JJ - (NV - II) * (NV - II + 1) / 2
```

The variable NREC could then be used in a disk READ or WRITE statement.

Accessing so many small records would, however, slow down a program, especially if access to every element in the upper triangle of the matrix is required for each subject during data input. An alternative way to define our disk file to hold a symmetric matrix of maximum dimension 200 would be as follows:

DEFINE FILE 7 (132,320,U,N7)

Using this configuration of the file, we can access elements of the matrix in blocks. To see how this works, consider the following method of accumulating a disk-resident version of the matrix of sums of squares and products described in Chapter 4.

First, we dimension an array X with 320 elements and zero-out the elements of our disk-resident matrix:

$$\text{DO } 20 \text{ I} = 1,320$$
$$20 \quad \text{X(I)} = 0.0$$
$$\text{NTOT} = (\text{NV} * \text{NV} - \text{NV}) / 2 + \text{NV}$$
$$\text{LAST} = \text{NTOT} - 1$$
$$\text{N7} = 1$$
$$\text{DO } 25 \text{ I} = 1,\text{LAST},320$$
$$25 \quad \text{WRITE } (7'\text{N7}) (\text{X(J)}, \text{J} = 1,320)$$

where N7 is the associated variable. The use of LAST as the upper limit of the index of the DO 25 loop ensures that no attempt to write a nonexistent record will occur.

Second, we input data for NS subjects and accumulate sums as shown in Figure 5-3. The arrays DATA and SUMX are each dimensioned 200 (the maximum number of variables), and SUMX is assumed to have been previously initialized to zero. Because the FIND statement (number 115) has been used, subroutine TAKE should not be LOCALed. The use of the array X to transfer 320 elements of the compressed matrix with each disk I/O statement greatly reduces the proportion of time spent in disk I/O. Efficiency could be improved still more if the dimension of X were set at a higher multiple of 160 (each 160 real-array elements occupy one record of file 7), since still fewer disk accesses would then be required.

Subsequent element-by-element access to our matrix could be obtained by using the LINK capability to transfer control to a mainline in which a new configuration of file 7 is created:

DEFINE FILE 7 (21120,2,U,N7)

This version of file 7 can be accessed by the method just described. The total of 21,120 two-word records is required to maintain a file with 132 * 320 words of storage. Be sure to read the later section, *Passing Temporary Files between Mainlines*, before attempting this strategy.

Printing of a disk-resident compressed symmetric matrix may be accomplished using either subroutine DCPTN (with variable names) or subroutine DCPT (without variable names).

The calling sequence for subroutine DCPT is

CALL DCPT (KFILE,N,LINES)

```
      C
      C      ACCUMULATION OF SUMS IN A DISK-RESIDENT COMPRESSED SYMMETRIC
      C         MATRIX
      C
             DO 115 I = 1,NS
      C
      C      INITIALIZE COUNTERS.
      C
             NE = 0
             NREC = 1
             NELS = 320
      C
      C      INPUT DATA FOR ONE SUBJECT.
      C
             CALL TAKE (KF,DATA,NV,KARDS,LINES)
      C
      C      READ FIRST BLOCK OF FILE.
      C
             READ (7'1) (X(J), J = 1,320)
      C
      C      BEGIN ACCUMULATION OF SUMS.
      C
             DO 110 J = 1,NV
             SUMX(J) = SUMX(J) + DATA(J)
             DO 110 K = J,NV
             NE = NE + 1
      C
      C      CHECK TO SEE IF ANOTHER BLOCK OF FILE IS NEEDED.
      C
             IF (NE - NELS) 105,105,100
      C
      C      RETURN BLOCK IN CORE TO DISK.
      C
         100 WRITE (7'1) (X(L), L = 1,320)
      C
      C      INCREMENT COUNTERS.
      C
             NREC = NREC + 1
             NELS = NELS + 320
      C
      C      READ NEXT BLOCK OF FILE.
      C
             READ (7'NREC) (X(L), L = 1,320)
      C
      C      DETERMINE WHICH BLOCK ELEMENT TO INCREMENT.
      C
         105 NBE = NE - NELS + 320
      C
      C      ADD PRODUCT TO SUM.
      C
         110 X(NBE) = X(NBE) + DATA(J) * DATA(K)
      C
      C      RETURN LAST BLOCK TO DISK.
      C
             WRITE (7'NREC) (X(J), J = 1,320)
      C
      C      LOCATE FIRST RECORD IN FILE.
      C
         115 FIND (7'1)
```

Figure 5-3. Accumulation of Sums in a Disk-Resident Compressed Symmetric Matrix

where   KFILE = number of disk file in which matrix is stored as a series of
                two-word records
         N = number of rows and columns in the matrix
       LINES = device number for the line printer
The matrix stored on disk is printed in eight-column blocks.
     The calling reference for subroutine DCPTN is

                 CALL DCPTN (KFILE,N,LINES,NFILE)

where KFILE, N, and LINES are defined as above, and NFILE is the number of
the disk file in which variable names are stored, 20 characters per name, one
name per record, one character per word. In this case, the matrix stored on disk
is printed in six-column blocks. Listings of both DCPT and DCPTN appear in
Appendix A.

DISK STORAGE OF A COMPRESSED NONSYMMETRIC MATRIX

As in the case of a symmetric matrix, the one-dimensional representation of a nonsymmetric matrix can be stored in a disk file of two-word records. For example, a data matrix of up to 20,000 elements total (i.e., NV * NS ≤ 20,000) could be stored in a file defined by the statement

DEFINE FILE 9 (20000,2,U,N9)

and loaded by the following statements:

```
        N9 = 1
        DO 10 I = 1,NS
        CALL TAKE (KF,X,NV,KARDS,LINES)
    10  WRITE (9'N9) (X(J), J = 1,NV)
```

where N9 is the associated variable and the dimension of X determines the maximum number of variables.

The number of the two-word record corresponding to element I,J of the uncompressed data matrix is given by the expression

NREC = (I - 1) * NV + J

When access by variables is required, as in the sequence

```
        DO 50 J = 1,NV
        CSUM(J) = 0.0
        DO 50 I = 1,NS
        NREC = (I - 1) * NV + J
        READ (9'NREC) ELEMT
    50  CSUM(J) = CSUM(J) + ELEMT
```

for computing column sums, considerable time will be consumed by disk activity. When access by subjects is required, as in the case of computing row sums, some improvement in run time can be achieved by reading all variables for one subject in a single disk READ:

```
        DO 55 I = 1,NS
        RSUM(I) = 0.0
        NREC = (I - 1) * NV + 1
        READ (9'NREC) (X(J), J = 1,NV)
        DO 55 J = 1,NV
    55  RSUM(I) = RSUM(I) + X(J)
```

Some form of blocking would, of course, result in still better performance, and the feasibility of blocking should be carefully examined in the context of the patterns of access to disk-resident array elements demanded by a particular application.

Modification of subroutine MPRT for printing a disk-resident compressed nonsymmetric matrix involves the following steps (referring to the listing of subroutine MPRT in Appendix A):

1. Remove X from the argument list and the DIMENSION statement.
2. Remove the statements between the DIMENSION statement and statement 40.
3. Add to the argument list NFILE, the number of the disk file in which the matrix is stored as a series of two-word records.
4. Change the third statement after statement 55 from KK = L + ND * (M - 1) to NREC = (L - 1) * NC + M.
5. Change statement 60 to 60 READ (NFILE'NREC) Y(LL).
6. Remove ND from the argument list.

Passing Temporary Files between Mainlines

When one mainline transfers control to another mainline by means of a CALL LINK statement, information may be transmitted via temporary files much in the same way that information is passed via COMMON. The procedure is really quite straightforward if some simple precautions are observed.

FILE ORDER

If several files are involved, they must be in the same order from mainline to mainline. Although file numbers are independent from mainline to mainline, like variable names in COMMON, it is wise to keep them the same.

FILE SIZE

Each file must occupy the same number of sectors from mainline to mainline. The easiest way to ensure this is to use the same set of DEFINE FILE specifications from mainline to mainline.

REDEFINITION OF FILE STRUCTURE

If care is taken to make sure a given file occupies the same number of sectors from one mainline to another, the configuration of the file can be different in different mainlines. For example, the statement

DEFINE FILE 6 (1000,160,U,N6)

could be replaced by the statement

DEFINE FILE 6 (500,320,U,N6)

in the corresponding position in another mainline.

IBM 1130 DISK MONITOR AND LIBRARY PROGRAMS

The reader of this section is assumed to be familiar with the basic IBM 1130 control cards. In the next few paragraphs we shall discuss a few monitor features and control cards of particular interest to the programmer of statistical library programs. This section is *not* intended to constitute a general introduction to the IBM 1130 operating system.

Storage of Programs

The mode in which library programs should be stored is dictated by a number of considerations that we shall examine in the context of a review of the control cards involved. Also included in the review of control cards will be implementation of the LOCAL feature discussed in Chapter 4. More extensive coverage of the IBM 1130 disk monitor system may be found in the *IBM 1130 Disk Monitor System, Version 2, Programming and Operator's Guide* (Form GC26-3717). Of course, while // JOB T should be used during debugging, // JOB must be used for permanent storage of either programs or files.

*STORE

As shown in Figure 5-4, this DUP (disk utility program) control record may be used to move a mainline, named MAIN, from working storage (WS) on

1	2	3	4	5	6	7	8	9	10	11	12	13	14	15	16	17	18	19	20	21	22	23	24	25	26	27	28	29	30	31	32	33	34	35	36	37	38	39	40	41
/	/		D	U	P																																			
*	S	T	O	R	E						W	S		U	A		M	A	I	N						0	0	0	1			0	0	0	2					

Figure 5-4. Example of *STORE Control Record

cartridge 0001 to the user area (UA) on cartridge 0002. If storage of the program on the system cartridge is desired, the cartridge ID's may be omitted. A program stored in this manner is in disk storage format (DSF).

EXECUTION OF A DSF PROGRAM

Figure 5-5 shows the control cards necessary to execute program MAINA, which references the LOCALed subprograms SUB1, SUB2, SUB3, and SUB4.

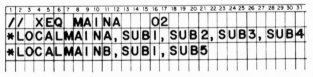

Figure 5-5. Execution of a DSF Program

Program MAINA subsequently transfers control (using CALL LINK) to DSF program MAINB, which references the LOCALed subprograms SUB1 and SUB5. The entry in columns 16 and 17 of the // XEQ card gives the number of supervisor control records that follow. If either program MAINA or program MAINB references permanent files, an appropriate entry for each file must appear on one or more *FILES cards (described below) included among the supervisor control records. If no supervisor control records are needed for either MAINA or MAINB, the card

// XEQ MAINA

would suffice.

STORECI

This DUP control record may be used to store a mainline program in disk core image format (DCI). The control records in Figure 5-6 would cause program

Figure 5-6. Example of a *STORECI Control Record

MAIN8, which references the LOCALed subprograms SUBA, SUBB, and SUBC, to be moved from working storage (WS) on cartridge 0001 to the user area (UA) on cartridge 0003. As in the case of the *STORE control record, references to cartridge ID's may be omitted if only the system cartridge is involved. The name of the mainline must be omitted from *LOCAL control record(s) used with *STORECI. The entry in columns 29 and 30 of the *STORECI card specifies the number of supervisor control records to be read (including *FILES cards, if any permanent files are to be referenced).

EXECUTION OF A DCI PROGRAM

Suppose the DCI program FAST terminates in a link to SLOW, a DSF program referencing the LOCALed subprograms ONE, TWO, THREE, and FOUR. Execution of program FAST (and, later, program SLOW) would require the control cards shown in Figure 5-7. If program FAST linked to programs requiring no supervisor control records, its execution would be begun by the control card

// XEQ FAST.

Figure 5-7. Execution of a DCI Program

REFERENCING A USER-SUPPLIED SUBPROGRAM AT EXECUTION TIME

The DSF programs can incorporate a user-supplied subprogram at execution time. Suppose, as depicted in Figure 5-8, the steps in computation of

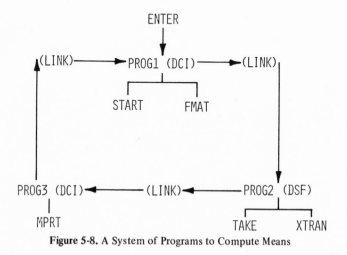

Figure 5-8. A System of Programs to Compute Means

means for NV variables measured for NS subjects have been divided among three mainline programs, each of which references one or more subprograms.

Stored in DCI, PROG1 inputs and echos control cards, defines parameters, zeros storage, and links to DSF PROG2. After reading data and accumulating ΣX for each variable, PROG2 links to DCI PROG3. PROG3 computes and prints the NV means and links to PROG1, which looks for possible additional data.

Figure 5-9 contains a listing of the source deck for PROG2. Values for KARDS, LINES, NV, and NS are transmitted via COMMON, and the elements of SUMX are assumed to have been set equal to 0.0 in PROG1. A listing of the

```
C
C       PROGRAM PROG2
C
        COMMON KARDS, LINES, NV, NS, X(200), SUMX(200), KF(630)
C
C       BEGIN INPUT OF DATA.
C
        DO 5 I = 1,NS
C
C       INPUT DATA FOR ONE SUBJECT.
C
        CALL TAKE (KF,X,NV,KARDS,LINES)
C
C       CALL USER-SUPPLIED TRANSFORMATION ROUTINE.
C
        CALL XTRAN (X)
C
C       ACCUMULATE SUMS.
C
        DO 5 J = 1,NV
      5 SUMX(J) = SUMX(J) + X(J)
C
C       TRANSFER CONTROL TO NEXT SEGMENT.
C
        CALL LINK (PROG3)
        END
```

Figure 5-9. Listing of Source Deck for Program 2

deck a user might submit for processing appears in Figure 5-10. The source deck for subroutine XTRAN appears as a module just before // XEQ PROG1, so XTRAN would be stored in DSF in the user area when PROG2 is loaded. The cards immediately following // XEQ PROG1 are program control cards read

```
// *
// *    SAMPLE JOB DECK FOR PROG1
// *
// JOB T
// FOR USE WITH PROG1
*ONE WORD INTEGERS
*LIST SOURCE PROGRAM
        SUBROUTINE XTRAN (X)
        DIMENSION X(200)
C
C       INSERT STATEMENTS NEXT.
C
        X(7) = SQRT(X(7))
        X(25) = X(25) - 10.5
        RETURN
        END
// DUP
*STORE        WS  UA  XTRAN
// XEQ PROG1
THIS IS A SAMPLE PROBLEM PREPARED FOR
ILLUSTRATION V-12.  THERE ARE 150
VARIABLES AND 200 SUBJECTS.
  150 200
 (10X, 70F1.0)
*** DATA FOR 200 SUBJECTS ***
*** BLANK CARD (STOPS PROGRAM) ***
*** BLANK CARD ***
```

Figure 5-10. Sample Job Deck for PROG1

by subroutines START and FMAT as called in PROG1: three title cards, a parameter card, and a format card. The first blank card after the 200 data cards causes subroutine START to branch to a CALL EXIT; the last blank card is required to permit the next-to-the-last card to be read by the IBM 1442 card READ/PUNCH. Because // JOB T is used, subroutine XTRAN is automatically deleted when the next JOB card is processed.

*STORE VERSUS *STORECI

Table 5-3 contains a brief summary of the comparative characteristics of programs stored by *STORE and *STORECI. The basic trade-off between *STORE and *STORECI is speed versus disk space. While a DSF program requires relatively little disk space, a delay of as much as several minutes occurs between the time the // XEQ card is read and actual execution begins. On the other hand, the quick-starting DCI program has all its subprograms built into itself, so a relatively large amount of disk space is required to store it.

Table 5-3. *STORE versus *STORECI

CONSIDERATION	*STORE	*STORECI
Delay after // XEQ card	Substantial	Negligible
Disk storage required	Minimal	Maximal
Control cards to execute	Can be complicated	Always simple
Reference to user subprograms	Possible	Impossible
Program maintenance	Straightforward	Cumbersome
Reference to permanent files	Straightforward	Complicated

As demonstrated above, LOCALs can require the inclusion of extra control cards after the // XEQ card if one or more DSF programs are to be executed. If only DCI programs (and/or DSF programs without LOCALs or references to permanent files) are to be executed, only a single // XEQ card is required.

If the application requires provision for the user to supply a subprogram for use at execution time, only a DSF program can fill the bill. Because, as mentioned above, the DCI program must have all its subprograms built in at the time it is stored, it is impossible to supply a subprogram to it at execution time. This same characteristic of the DCI program complicates program maintenance. If a new version of a subprogram is to be incorporated into a DCI program, the

DCI program must be deleted and rebuilt (re-stored, using *STORECI). This is, of course, not true for a DSF program.

Finally, for reasons we shall discover in the discussion of permanent files below, reference to permanent files is much less of a problem with DSF programs than with DCI programs. Temporary files pose no particular problem for either mode of storage, but reference to permanent files involves more complications using *STORECI than using *STORE.

Watch for SOCALs

SOCALs (system-overlays-to-be-loaded-on-call) are system subprogram groups made into overlays by the operating system. When a mainline program and its subprograms will not fit into core memory, the system will automatically attempt to reduce the core memory requirements of the core load by using SOCALs. This tactic often works, and the program is successfully loaded.

Because SOCALs involve subprograms for such operations as real arithmetic and disk FORTRAN I/O, they can exact a high cost in decreased execution speed. Accordingly, care should be taken to make sure a program is not SOCALing.

Messages to the effect that SOCALs are required are printed automatically at the time a *STORECI is performed. If a DSF program or a program in working storage is to be executed, a core map and associated messages concerning any SOCALs can be obtained by punching an L in column 14 of the // XEQ card. If SOCALs are discovered in any case, it usually is advisable to modify the program (through segmentation or a decrease in array dimensions) to eliminate them.

USE OF DISK FOR PERMANENT FILES

The few statistical program library users who are familiar with FORTRAN and also possess substantial amounts of data may want to examine the possibility of creating permanent disk-resident data files for subsequent input to various library programs. This approach is particularly attractive at IBM 1130 installations featuring multiple disk drives.

Creating a Permanent File

Because most of the techniques previously discussed in connection with temporary files also apply to permanent files, we shall approach the creation of a permanent file in terms of the steps required to make a temporary file permanent.

DETERMINING SECTOR COUNT

Disk space for permanent files is allocated in sectors, and the first thing we must do is determine the total number of 320-word sectors required for our file. Our task is complicated by the fact that records may not extend across sector boundaries. This means that certain record lengths will cause a part of each sector to be unused. For example, a file defined by the statement

$$\text{DEFINE FILE 8 } (1000,57,U,N8)$$

would result in allocation of five records per sector with 35 words per sector unused.

From our DEFINE FILE statement we obtain NWR, the number of words per record, and NR, the number of records. Then, using the syntax of FORTRAN (but computing by hand), we can compute NSF, the number of sectors of disk storage required for our file:

$$NRS = 320 \,/\, NWR$$
$$NSF = NR \,/\, NRS$$
$$IF \, (NR - NRS * NSF) \, 10,10,5$$
$$5 \quad NSF = NSF + 1$$
$$10 \quad (done)$$

For example,

$$\text{DEFINE FILE 7 } (1505,150,U,N7)$$

would yield NR = 1505 and NWR = 150. The variable NRS would be 2, and the division of 1505 by 2 would give us 752 and a nonzero remainder, so we would round up to 753 as a final value for NSF.

ESTABLISHING A PERMANENT AREA ON DISK

Next we use a DUP control record to set up a named area on a disk consisting of NSF sectors. This can be done one of two ways.

As portrayed in Figure 5-11, the *STOREDATA operation can be used to set up the area. In our example, permanent file PERM1 consisting of 753 sectors is established in the user area of cartridge 0099. Working storage is located on cartridge 0001. Cartridge ID's may be omitted if the file is to be on the system cartridge.

If working storage is smaller than NSF sectors, *STOREDATA will not

Figure 5-11. Use of *STOREDATA to Create a Permanent File

work, since NSF sectors must be moved from working storage. On systems operating under Version 2, modification 8 or higher, of the IBM 1130 disk monitor
system, the *DFILE operation shown in Figure 5-12 may be used to establish a
permanent file. Because *DFILE circumvents working storage, it is not vulnerable to the problem described above. The example in Figure 5-12 parallels the
earlier *STOREDATA.

Figure 5-12. Use of *DFILE to Create a Permanent File

*FILES CARD

At this point, the disk area created by *STOREDATA or *DFILE is un-
defined. We now need a means of associating the file defined in our DEFINE
FILE to the disk area we have just created. The *FILES card provides this
capability, so we can load our file with data using the program in Figure 5-13.
If it were not for the *FILES card, file 7 would reside in working storage as a
temporary file. The *FILES card permits us to establish an equivalence between
the logical file number defined within our program and the named area on disk.

The program presented in Figure 5-13 has been simplified for the sake of
clarity. In actual practice we would want to use many of the blocking techniques
discussed in the introductory section "Use of Disk for Temporary Files."

Interfacing Permanent Files and Library Programs

Because any required *FILES card must follow the *STORECI card at the
time a DCI program is stored on disk, it would at first appear that only a DSF
program (whose // XEQ card may be followed by supervisor control records,
including a *FILES card) could access a permanent file. While a library of DSF
programs could be used for direct interface with permanent files, there is a
method for a DCI program to access a permanent file indirectly.

The secret is simple: In a user-written mainline program, data to be processed by the library program are transferred to a standard temporary file. The

```
// *
// *   LCADING A PERMANENT FILE
// *
// JOB    0001 0099
// FCR
*ONE WCRD INTEGERS
*LIST SOURCE PROGRAM
*IOCS(CARC,DISK,1403PRINTER)
        CIMENSION ID(6), X(72)
        CEFINE FILE 7 (1505,150,U,N7)
C
C       BEGIN INPUT.
C
        CO 10 I = 1,1505
C
C       REAC ONE SUBJECT'S DATA.
C
        READ (2,5) (ID(J), J = 1,6), (X(K), K = 1,72)
      5 FORMAT (6A2, 34F2.0 / 12X, 34F2.0 / 12X, 4F3.0)
C
C       STCRE ONE SUBJECT'S DATA ON CISK.
C
     10 WRITE (7'I) (IC(J), J = 1,6), (X(K), K = 1,72)
C
C       READ DATA BACK FROM CISK AND PRINT ECHO CHECKS.
C
        CC 15 I = 1,1505
        READ (7'I) (ID(J), J = 1,6), (X(K), K = 1,72)
     15 WRITE (5,20) (ID(J), J = 1,6), (X(K), K = 1,72)
     20 FORMAT (1X, 6A2, 18F4.0 / 3(13X, 18F4.0))
        CALL EXIT
        END
// XEC        01
*FILES(7,PERM1,0099)
(BLANK CARD)
```

Figure 5-13. Loading a Permanent File

statement CALL LINK is then used to transfer control to the library DCI program, which, upon recognition of a nonzero entry in the appropriate parameter card field, reads its input data from the standard temporary file rather than cards. An example of the use of this method may be found in Chapter 3 in the special write-up for disk input to program C01. Program C01 itself is presented in detail in Chapter 8.

SUMMARY

In this chapter we have considered the use of the IBM 1130 disk for storage of temporary data files, permanent data files, and programs. Included in all discussions of data files was the concept of blocking information so that only occasional disk access is required. In the section on program storage, particular attention was paid to a comparison of the implications of the use of the *STORE and *STORECI control records.

It is hoped that the suggestions in this chapter will help IBM 1130 users in more effectively exploiting the disk in implementations of statistical procedures.

6 Use of IBM 1130 Commercial Subroutine Package

In the days when only assembly language and FORTRAN were available for the IBM 1130, the IBM 1130 commercial subroutine package (CSP) was developed to provide capabilities needed in business (commercial) applications. The CSP includes a routine for decoding input records after they have been read, an editing routine for the preparation of output in special formats, code conversion routines for data manipulation and more efficient data packing, routines for variable-length decimal arithmetic, routines for improved speed and control of I/O devices, and several utility routines for common tasks in a commercial programming environment.

Not all CSP routines are of great interest to the programmer of statistical library programs. In this chapter we shall consider only a limited subset of CSP routines judged to be of particular usefulness in statistical applications. A more thorough introduction to the CSP from a commercial applications frame of reference may be found in *1130 Commercial Subroutine Package (1130-SE-25X), Version 3 Program Reference Manual* (Form H20-0241) or Chapter 10 in Louden (1967).*

CSP ROUTINES FOR INPUT/OUTPUT

The common ground for statistical routines and the CSP is processing of information represented as a series of characters stored in elements of one or

---

*R. K. Louden, *Programming the IBM 1130 and 1800*, Englewood Cliffs, N. J., Prentice-Hall, Inc., 1967.

more one-dimensional integer arrays. For example, at several points in our discussion of program control cards, we have advocated card input in the form of a character string:

$$\text{READ (KARDS,50) (KARD(I), I = 1,80)}$$
$$50 \quad \text{FORMAT (80A1)}$$

Most of the routines of the CSP are written to handle information stored in this form.

Card Input

SUBROUTINE READ

On an IBM 1130 with an IBM 1442 card READ/PUNCH, the statement

$$\text{CALL READ (KARD,1,80,NER)}$$

is equivalent to the regular FORTRAN READ and FORMAT given just above; however, it is a considerably faster means of reading a card as 80 characters. The increase in speed is due to the fact that subroutine READ is an assembly language routine that overlaps card input with the conversion of card codes to A1 format.

The general form of the calling sequence of subroutine READ is

$$\text{CALL READ (KARD,NSTAR,NSTOP,NER)}$$

where   KARD = one-dimensional integer array into which the contents of a card will be read, in A1 format, one character per element
    NSTAR = first element of KARD into which a character will be read
    NSTOP = last element of KARD into which a character will be read
    NER = condition indicator

The integer variable NER must be initialized to -1 before the call to subroutine READ, and is returned containing 0 (zero) if the last card has been physically read into core memory or is returned as 1 if a feed or read check has occurred during the attempt to read the card. After a call to subroutine READ, NER may be tested:

$$\text{NER} = -1$$
$$25 \quad \text{CALL READ (KARD,1,80,NER)}$$
$$\text{IF (NER) 30,50,25}$$
$$30 \quad \text{(normal processing)}$$
$$\cdot$$
$$\cdot$$
$$\cdot$$
$$50 \quad \text{(last card has been encountered)}$$

In this example, the branch back to statement 25 permits recovery from a read or feed check.

Because the information returned in NER is not particularly useful in the stacked-job, open-shop environment in which we assume statistical library programs will be used, no further discussion of its use is included here. In the uses of subroutine READ described in this chapter, any available integer variable should be supplied for NER.

Although most of our calls to subroutine READ will feature values of 1 and 80 for NSTAR and NSTOP, respectively, other values may be used (subject to the restriction that NSTOP − NSTAR + 1 ⩽ 80). For example, the statement

CALL READ (KARD,81,160,NER)

would place the contents of a card in elements 81 through 160 of KARD, and the statement

CALL READ (KFEW,26,50,NER)

would place the contents of columns 1 to 25 of a card in elements 26 through 50 of KFEW. Input of characters always begins with column 1 of the card, and only one card may be read in a single call to subroutine READ.

Users of an IBM 1130 with an IBM 2501 card reader may call subroutine R2501, which features a calling sequence exactly the same as that for subroutine READ. Although subroutine READ is featured in the examples in this chapter, the same remarks apply to subroutine R2501 unless otherwise noted.

SPECIAL CHARACTERISTICS OF SUBROUTINE READ

Unlike an ordinary FORTRAN READ statement referencing logical unit 2, subroutine READ does not detect monitor (//) control cards. Accordingly, if such detection is desired, tests must be included in the input sequence:

```
        DIMENSION MON(3), KARD(80)
        DATA MON /'/','/',' '/
        .
        .
        .
        CALL READ (KARD,1,80,NER)
        DO 5 I = 1,3
        IF (KARD(I) - MON(I)) 10,5,10
     5  CONTINUE
        (statements to respond to detection of monitor control card)
    10  (normal processing)
```

Subroutine READ also handles unrecognized characters differently from an ordinary FORTRAN READ statement. For example, the statements

READ (KARDS,5) (KARD(I), I = 1,80)
5 FORMAT (80A1)

will generate an F-error if certain characters (such as ';', a semicolon) present on the IBM 029 keypunch, but unrecognized by FORTRAN, are encountered on the card being read. Subroutine READ, on the other hand, will simply process the non-FORTRAN character without protest, placing the appropriate EBCDIC code in the one-dimensional integer array specified in the CALL statement. A decision concerning the handling of the code can be made later.

SUBROUTINE TAKEC

Subject to the restrictions discussed under a subsequent heading, "Restrictions Concerning CSP Input/Output," a call to subroutine READ may be used anywhere that input of a card as a character string is desired. Where the simulation of variable format is to be used, a modified version of subroutine TAKE should be called. This version, which calls subroutine READ, has the following calling order:

CALL TAKEC (KF,X,NV,LINES)

Here, as in the calling order for subroutine TAKE,

KF = one-dimensional integer array containing format specifications encoded by a call to subroutine FMAT

X = one-dimensional real array in which scores are to be placed, beginning with X(1)

NV = number of scores to be read

LINES = device number for the line printer

Thus, although subroutine TAKE has been included in discussions of data input throughout previous chapters of this book, IBM 1130 users should substitute the faster subroutine TAKEC with its slightly different argument list.

Card Output

SUBROUTINE PUNCH

Because the IBM 1442 card READ/PUNCH features only one card path passing through both read and punch stations, it is desirable to take precautions to avoid inadvertently punching on a nonblank card in a stacked-job environment:

```
        DIMENSION KARD(80)
        DATA KBLNK/' '/
        .
        .
        .
    45  READ (2,50) (KARD(I), I = 1,80)
    50  FORMAT (80A1)
        DO 60 I = 1,80
        IF (KARD(I) - KBLNK) 55,60,55
    55  (statements to cope with nonblank card)
    60  CONTINUE
        WRITE (2,65) list
    65  FORMAT (...
```

Because statements for regular FORTRAN I/O and CSP I/O for the same device cannot appear in the same core load (see section "Restrictions Concerning CSP Input/Output"), it is not permissible to replace statements 45 and 50 with a call to subroutine READ and leave the WRITE statement and its FORMAT in the program. Instead, punched output must be generated by a call to subroutine PUNCH, which has the following calling sequence:

$$\text{CALL PUNCH (KK,NSTAR,NSTOP,NER)}$$

where   KK = one-dimensional integer array containing the characters to be punched, in A1 format, one character per element

NSTAR = first element of KK to be punched

NSTOP = last element of KK to be punched

NER = condition indicator

As in the case of subroutine READ, we shall ignore NER. Also, as in the case of subroutine READ, any values may be used for NSTAR and NSTOP, subject to the restriction that NSTAR $\leqslant$ NSTOP and NSTOP $-$ NSTAR $+ 1 \leqslant 80$. Elements of KK are always punched, beginning in column 1 of the card.

SUBROUTINE ECODE

While subroutine PUNCH is straightforward in its calling sequence and operation, the preparation of the string of characters for it to punch is often quite cumbersome. Suppose, for example, our WRITE and FORMAT to be replaced are as follows:

```
        WRITE (2,90) (ID(J), J = 1,12), NTOT
    90  FORMAT (12A1, 5X, I5)
```

We must prepare a 22-element array KK with the information previously represented in the list of our WRITE statement. We begin by copying the array ID into KK:

$$DO\ 90\ J = 1,12$$
$$90\quad KK(J) = ID(J)$$

Then, using the statement

$$DATA\ KBLNK\ /'\ '/$$

to set up the integer equivalent of a blank, we blank out the elements of KK corresponding to the 5X in our FORMAT statement:

$$DO\ 95\ J = 13,17$$
$$95\quad KK(J) = KBLNK$$

So far, so good, but what should we do with NTOT? Fortunately, a subroutine (not part of the CSP; a listing appears in Appendix A) is available to help us:

$$CALL\ ECODE\ (KK,NSTAR,NSTOP,INTEG)$$

where   KK = one-dimensional integer array into which INTEG is to be placed as characters, one digit per element

NSTAR = first element of KK to be filled

NSTOP = last element of KK to be filled

INTEG = number to be converted to characters and placed in KK

In other words, when KK is an 80-element array representing a card, NSTAR and NSTOP define the left- and right-hand limits of an output field. The string of characters (numerals) representing INTEG are right-justified in the field, and if INTEG is negative, a minus sign is inserted in front of the first numeral. Any unused portion of the field is blanked out.

The variable NTOT from the list of our WRITE statement can now be inserted in the output array KK:

$$CALL\ ECODE\ (KK,18,22,NTOT)$$

The single statement

$$CALL\ PUNCH\ (KK,1,22,NER)$$

then punches the card.

SUBROUTINE PUNCH VERSUS WRITE

If real variables appear in the list of a WRITE statement referencing the IBM 1442 card READ/PUNCH, the cumbersomeness of subroutine PUNCH is increased. While a CSP routine, subroutine PUT, is available to place a real variable in an output array, the core load must be in extended precision. If extended precision is not desired, an older, less accurate FORTRAN version of subroutine PUT may be found in Louden.*

In short, unless punched output in a core load including references to subroutine READ is essential, the standard FORTRAN WRITE statement is to be favored over subroutine PUNCH.

ALTERNATE STACKER SELECTION

Although the IBM 1442, Model 6 or 7, has two stackers, only one is usually used in a FORTRAN environment. Subroutine STACK is an assembly language CSP routine that allows the FORTRAN programmer to divert cards into the alternate stacker.

Suppose, for example, there is a need to perform an analysis using only data from the males in a sample of 1025 subjects. The program presented in Figure 6-1 is designed to read 1025 sets of three cards each, making sure an identification field in columns 1 through 10 is the same for each set of three cards and that cards 1, 2, and 3 are present and in order for each subject. Using subroutine STACK, the program deposits in the alternate stacker all cards identified as representing male subjects.

As can be seen by an inspection of statement 35, subroutine STACK has no arguments. When it is called, it simply selects the alternate stacker for the next card to go through the punch station. Because the punch station follows the read station, the card READ (using subroutine READ) just before the call to subroutine STACK is the card selected.

The division of the counter MALES by 3 before printing the total number of males selected is necessitated by the fact that MALES is incremented each time a card is selected, and there are three cards per subject.

The program in Figure 6-1 has been simplified for clarity, and it is assumed that the reader will readily imagine applications of subroutine STACK more appropriate to his particular operating environment.

Restrictions Concerning CSP Input/Output

First, programs referencing *any* CSP routine (not just I/O routines) must be compiled with the *ONE WORD INTEGERS control statement.

Second, with the exception of subroutine STACK, CSP I/O and regular

---

*R. K. Louden, *Programming the IBM 1130 and 1800*, Englewood Cliffs, N. J., Prentice-Hall, Inc., 1967.

```
C
C      PROGRAM TO REMOVE MALES FROM A DATA DECK
C
       DIMENSION KARD(12), ID(10)
C
C      INITIALIZE COUNTER.
C
       MALES = 0
C
C      BEGIN INPUT OF DATA.
C
       DO 60 I = 1,1025
C
C      INPUT FIRST CARD OF ONE SUBJECT'S SET.
C
       CALL READ (KARD,1,12,NER)
C
C      COPY SUBJECT IDENTIFICATION INTO ID.
C
       DO 5 J = 1,10
     5 ID(J) = KARD(J)
C
C      SET CARD NUMBER INDICATOR.
C
       J = 1
C
C      DECODE AND CHECK CARD NUMBER.
C
    10 CALL DCODE (KARD,12,12,NUM,IER)
       IF (IER) 15,15,50
    15 IF (NUM - J) 50,20,50
C
C      MAKE SURE SUBJECT ID MATCHES FIRST CARD OF SET.
C
    20 DO 25 K = 1,10
       IF (KARD(K) - ID(K)) 50,25,50
    25 CONTINUE
C
C      DECODE AND CHECK SEX (1 = MALE).
C
       CALL DCODE (KARD,11,11,NUM,IER)
       IF (IER) 30,30,50
    30 IF (NUM - 1) 40,35,40
C
C      SELECT CARD.
C
    35 CALL STACK
C
C      INCREMENT MALES * 3 COUNTER.
C
       MALES = MALES + 1
C
C      SEE IF ANOTHER CARD OF THIS SET IS TO BE READ.
C
    40 IF (J - 3) 45,60,60
C
C      READ NEXT CARD IN SET.
C
    45 CALL READ (KARD,1,12,NER)
C
C      INCREMENT CARD NUMBER INDICATOR AND RETURN TO CHECKING SEQUENCE.
C
       J = J + 1
       GO TO 10
C
C      PRINT ERROR MESSAGE.
C
    50 WRITE (1,55) (ID(J), J = 1,10)
    55 FORMAT (14HERROR AT ID = , 10A1)
       GO TO 65
    60 CONTINUE
C
C      PRINT TOTAL N SELECTED AND EXIT.
C
    65 MALES = MALES / 3
       WRITE (1,70) MALES
    70 FORMAT (I5, 16H MALES SELECTED.)
       CALL EXIT
       END
```

Figure 6-1. Program to Remove Males from a Data Check

FORTRAN I/O may not be intermixed in the same core load for the same device. This means, for example, that the statement

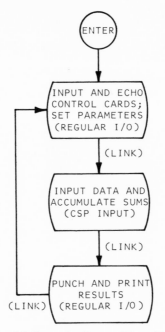

Figure 6-2. Separating Forms of Input/Output

READ (2,5) *list*

and

CALL READ (KK,1,25,NER)   or   CALL PUNCH (KK,2,15,NER)

may not appear in the same core load. There is, however, a simple way to get around this restriction. As shown in Figure 6-2, it is possible to segregate CSP I/O and regular FORTRAN I/O in separate mainlines that link to one another. In this way, any given core load (i.e., a mainline and its subprograms) uses only one form of I/O; yet CSP I/O can be used when particularly advantageous, as in reading a large data deck.

Since the *IOCS card contains references to only those devices for which regular FORTRAN I/O is used, a device for which CSP I/O is used is omitted from the IOCS card.

PACKING ALPHANUMERIC INFORMATION

Although most of the routines of the CSP are designed to operate on information stored in A1 format (one character per one-dimensional integer array element), CSP includes routines for converting from A1 format to other more efficient modes of representation of information. The user of CSP can thus rep-

resent information in A1 format for relative convenience when character manipulation is required, and can convert to a more compact form of storage when
core or disk memory is at a premium.

The other modes of representation available are A2, two characters per
one-dimensional integer array element, and A3, three characters (of a somewhat
restricted character set) per one-dimensional integer array element.

A2 Format

In regular IBM 1130 FORTRAN I/O, it is possible to input alphanumeric
information in the first 20 columns of a card into the elements of a one-
dimensional integer array in one of two ways:

$$5 \quad \text{READ } (2,10) \, (KK(I), I = 1,20)$$
$$10 \quad \text{FORMAT } (20A1)$$

or

$$15 \quad \text{READ } (2,20) \, (LL(I), I = 1,10)$$
$$20 \quad \text{FORMAT } (10A2)$$

The first form, A1 format, is the form associated with subroutine READ. The
second form, A2 format, provides storage of the same number of characters
in one-half of the core or disk memory.

SUBROUTINE PACK

Conversion from A1 format to A2 format is provided by subroutine
PACK, whose calling sequence is

CALL PACK (KK,KSTAR,KSTOP,LL,LSTAR)

where

KK = one-dimensional integer array containing alphanumeric information
in A1 format, one character per element
KSTAR = the first element of KK to be packed
KSTOP = the last element of KK to be packed
LL = a one-dimensional integer array into which the information from KK
is to be packed, in A2 format, two characters per element
LSTAR = the first element of LL to receive the packed characters.

An even number of characters is always taken from KK, even if KSTOP −
KSTAR + 1 is odd. If necessary, KK(KSTOP+1) is taken to make the total
number of characters even.

We can now produce the CSP equivalent of statements 15 and 20 of regular FORTRAN I/O given above:

CALL READ (KK,1,20,NER)
CALL PACK (KK,1,20,LL,1)

After execution of the preceding sequence, LL would contain two characters in each of its first ten elements.

SUBROUTINE UNPAC

A CSP routine to convert from A2 to A1 format has the following calling sequence:

CALL UNPAC (LL,LSTAR,LSTOP,KK,KSTAR)

where   LL = one-dimensional integer array containing alphanumeric information in A2 format, two characters per element
      LSTAR = first element of LL to be unpacked
      LSTOP = last element of LL to be unpacked
          KK = one-dimensional integer array into which the information from LL is to be unpacked, in A1 format, one character per element
      KSTAR = first element of KK to receive the unpacked characters
Two elements in KK will be required for each element in LL.

In practice, UNPAC would typically be used in moving information from storage into areas for active manipulation. For example, one might read a 40-word, A2 format disk record and unpack it for manipulation:

READ (99'I) (KTEMP(J), J = 1,40)
CALL UNPAC (KTEMP,1,40,KK,1)

Execution of these two statements would fill the first 80 elements of KK with information in A1 format.

A3 Format

The designation given this form of storage is a departure from standard FORTRAN terminology, since the IBM 1130 FORTRAN statements

READ (2,5) (KK(I), I = 1,10)
5   FORMAT (10A3)

would *not* successfully input the first 30 columns of a card. As used in connection with the CSP, the term "A3 format" does refer to the storage of three characters per integer array element; however, the programmer using A3 format must be willing to live with a character set consisting of only 40 characters. The set of 40 characters is defined by the programmer, so he can choose the 40 characters most appropriate to his application. The character set is defined by inserting the 40 characters (including a blank, if desired) in a one-dimensional integer array (which we shall call ICHAR). The characters should appear in the order of their most frequent occurrence, and the following 40 characters constitute a reasonable working set in the order shown:

0123456789ETAOINbSHRDLUCMFWYPVBGKQJXZ,.-

The "b" represents a blank. The array ICHAR may be read from a card or defined in a DATA statement.

SUBROUTINE A1A3

Packing in A3 format is accomplished by a call to subroutine A1A3:

CALL A1A3 (KK,KSTAR,KSTOP,LL,LSTAR,ICHAR)

where   KK = one-dimensional integer array containing alphanumeric information in A1 format, one character per element

KSTAR = first element of KK to be packed

KSTOP = last element of KK to be packed

  LL = one-dimensional integer array into which the information from KK is to be packed, in A3 format, three characters per element

LSTAR = first element of LL to receive the packed characters

ICHAR = array of characters described above

The number of elements of KK to be packed should be a multiple of 3. If a character in KK does not appear in ICHAR, it will be converted to a blank.

Using subroutine A1A3, information from three cards could be stored in one 80-element array:

CALL READ (KK,1,80,NER)
CALL READ (KK,81,160,NER)
CALL READ (KK,161,240,NER)
CALL A1A3 (KK,1,240,LL,1,ICHAR)

It is assumed that the elements of ICHAR have been previously defined. The array LL could subsequently be stored on disk in an 80-word record.

SUBROUTINE A3A1

Information previously packed by a call to subroutine A1A3 can be unpacked by a call to subroutine A3A1:

CALL A3A1 (LL,LSTAR,LSTOP,KK,KSTAR,ICHAR)

where   LL = one-dimensional integer array containing alphanumeric in-
                 formation in A3 format, three characters per element
         LSTAR = first element of LL to be unpacked
         LSTOP = last element of LL to be unpacked
            KK = one-dimensional integer array into which the information
                 from LL is to be unpacked, in A1 format, one character per
                 element
         KSTAR = first element of KK to receive the unpacked characters
         ICHAR = array of characters described above
Each element of LL yields three elements in KK.

We may reverse the process described in connection with subroutine A1A3, and retrieve the information from our three cards:

READ (101′I) (LL(J), J = 1,80)
CALL A3A1 (LL,1,80,KK,1,ICHAR)

Assuming that ICHAR has been previously defined as required, KK would contain 240 characters of alphanumeric information in A1 format after execution of these two statements. Of course the elements of ICHAR must be the same as those used with subroutine A1A3.

While character-oriented disk files are commonplace in commercial applications using the IBM 1130, disk files of data used in connection with statistical analyses tend to feature real or integer representation of numeric data. Accordingly, subroutines A1A3 and A3A1 are most useful for storage of such information as verbal responses or alphanumeric subject identification fields. For example, a 12-column subject identification field and associated data fields could be stored using the following statements:

READ (2,5) (ID(J), J = 1,12), (IT(K), K = 1,16)
5   FORMAT (12A1, 16I4)
CALL A1A3 (ID,1,12,IDP,1,ICHAR)
WRITE (7′I) (IDP(J), J = 1,4), (IT(K), K = 1,16)

File 7 has as many 20-word records as there are subjects; ID and IDP are dimensioned 12 and 4, respectively; and the 40-element array ICHAR has been defined in a DATA statement. The use of the integer array IT to store the 16 four-digit scores for each subject permits storage of the scores in one-half of the memory a real array of the same dimension would require.

A2 Format versus A3 Format

Assuming that a somewhat restricted character set of 40 characters (versus 48 in the unrestricted IBM 1130 FORTRAN set) is tolerable, what are the disadvantages of A3 format? In considering this question and the routines for conversion between A1 format and the two packed formats, we must realize that we are dealing with only two subprograms containing two entry points each. Subroutines PACK and UNPAC are actually two entry points to one routine, and subroutines A1A3 and A3A1 are two entry points to another routine. With this in mind, we can examine the two competing subprograms.

SPEED OF CONVERSION BETWEEN FORMATS

As can be seen in Table 6-1, the PACK/UNPAC routine is generally faster than the A1A3/A3A1 routine. For example, the unpacking of 3000 cards of information from A2 format to A1 format would require approximately 49 seconds if 1000 calls to subroutine UNPAC were involved. On the other hand, subroutine A3A1 would require approximately 114 seconds to do the same job.

Table 6-1. Conversion Speeds and Core Requirements

CONVERSION	EXECUTION TIME*	MEMORY†
A1 = A2	$360 + 63A$‡	66
A2 = A1	$420 + 66A$	
A1 = A3	$470 + 1084A$	192§
A3 = A1	$545 + 156A$	

*Approximate time in microseconds, assuming 3.6-microsecond central processing unit cycle speed.

†In words.

‡A = number of A1 characters.

§Including 40 words for array ICHAR.

SOURCE: *IBM 1130 Commercial Subroutine Package (1130-SE-25X), Version 3, Program Reference Manual* (Form H20-0241), White Plains, N. Y.: IBM, 1968.

CORE REQUIREMENTS

Although the A1A3/A3A1 routine is more than three times as large as the PACK/UNPAC routine, the difference of only 126 words is hardly likely to be critical in most situations. Neither routine requires a large amount of core memory.

AWKWARDNESS OF CONVERSION RATIOS

While it is no problem to pack a single card in A2 format, 80 is not a convenient number to divide by 3. One way to overcome this problem is to block information to be stored in multiples of three characters. In the case of cards, groups of three cards could be conveniently handled. It is true, however, that the 3:1 ratio of A3 format is generally more awkward than the 2:1 ratio of A2 format.

CONVENIENCE

Setting up the required 40-element array ICHAR for the A1A3/A3A1 routine is a minor annoyance that should be considered in choosing between the two modes of packing. The PACK/UNPAC routine requires no such array.

CONCLUSION

While no major difficulties are involved in the use of either A2 or A3 format, the somewhat decreased efficiency and programming ease associated with A3 format would seem to suggest its use only in applications in which the greater compactness of storage is really required.

SUMMARY

While the IBM 1130 commercial subroutine package (CSP) was developed for use in business-oriented programming in IBM 1130 FORTRAN, some of its routines are also quite useful in programming the statistical library. In particular, routines for input/output and packing of alphanumeric information can be put to good use in portions of library routines where character manipulation is involved.

The intent of this chapter has been to present a brief introduction to these useful CSP routines and to illustrate their use outside the commercial environment.

Although the CSP was first developed for the IBM 1130, other vendors (e.g., Digital Equipment Corporation and Hewlett-Packard) offer similar routines adapted for use on their systems.

7 Library Program Standards

In this chapter we consider a formalization of the structure associated with statistical library programs.* Although the writing of a computer program may be regarded as an act of self-expression, the creation of a library program should be an act involving rigorous self-discipline. The need for a set of standards arises from the context in which library programs are written—a community of researchers. Part of any harmonious community is a spirit of free exchange of ideas and resources, and computer programs constitute a vital community resource. Accordingly, statistical routines should be written in such a way that free exchange of program libraries is facilitated.

Another less colleagueship-minded reason for striving to impose standards upon library programs is the facilitation of training and communication within the staff of a single computer center. Programmers come and go, consuming valuable time as they struggle to attain familiarity with established practices and often leaving behind a collection of cryptic code. Attention to a set of formal standards can improve the utilization of programmers.

Because it is anticipated that frequent reference to the contents of this chapter may be required, the library program standards proposed are presented in outline form; however, as in many endeavors, it is probably less important to adopt any *particular* set of standards than it is to adopt *some* set of standard

---

*A number of the suggestions included in this chapter originated in discussions with the staff of the Vanderbilt University Computer Center, Nashville, Tenn.

practices. For this reason, the recommendations presented below should be treated as a model to be adapted to the needs of a particular computing environment.

INPUT STANDARDIZATION

1.0.0 *The problem deck setup should be as similar as possible for all programs in the library.*

The need for this general practice has been discussed in detail in previous chapters; indeed, the focus of Chapter 2 is on program control cards. It should, however, be noted that this recommendation applies to *all* control cards.

1.1.0 *The configuration of required system control cards should be as similar as possible for all programs in the library.*

System control cards are probably the most bewildering aspect of library programs. Because the typical program library user has had little or no programming experience and has no notion of the details of a computer's operating system, the format, order, and content of system control cards appear totally arbitrary to him. The user's discomfort is further compounded by the tendency for his errors in the preparation of system control cards to yield incomprehensible error messages (or no error messages at all). While the author of a library program can produce a routine that will diagnose errors with respect to program control cards and provide intelligible error messages, he has virtually no control over the error messages of the operating system. Any reduction of operating environment ambiguity must be achieved through clever program design.

1.1.1 *Library programs should be designed so that as few system control cards as possible are required.*

One way to minimize the likelihood of difficulty with system control cards is to minimize the number of opportunities for the user to commit an error. Standard device assignments should be used, for example, so that a system control card for device assignment need not appear in the job deck. On some systems the set of system control cards required for execution of a particular program can be stored in a special file. In this way a relatively complex set of system control cards could be "invoked" by only one or two simple system control cards. Still another way to minimize the required number of system control cards is storage of library routines as executable files (e.g., on the IBM

1130, stored by *STORECI). With the overhead associated with program loading out of the way, this mode of storage provides the added advantage of greatly increased throughput.

1.2.0 *The configuration of required program control cards should be as similar as possible for all programs in the library.*

Specific programs will, of course, require idiosyncratic control cards for the implementation of a particular statistical technique; nevertheless, certain control cards can be made a part of almost all programs in a library.

1.2.1 *The order of program control cards should be as similar as possible for all programs in the library.*

To some extent, use of the subprograms described in this book determines program control card order. For example, subroutine START looks for title cards before a parameter card. In particular, the following general card order has been recommended in Chapter 2:

1. Title Card(s) (Optional)
2. Parameter Card
3. Format Card(s) *or* Variable Location Card(s)
4. Variable Names Card(s) (Optional)
5. Missing Data Signal Card(s) (Optional)
6. Data Cards
7. Blank Card

If other program control cards are used in connection with several programs of the library, they should appear in a consistent location within the general card order.

1.2.2 *The layout of a particular control card should be consistent for all programs in the library.*

An obvious example of the application of this recommendation would be the case of the variable names card. As described in previous chapters, the variable names card is assumed to contain four 20-column fields whenever it is used with any program.

In addition to containing 20 four-column fields, a further standardization is suggested for the parameter card. Although any problem parameters may be inputted in any order via the parameter card, it is recommended that parameters associated with input (e.g., number of variables, number of subjects, or input options) be grouped together in the first fields of the card and that parameters associated with output (e.g., output options) appear last.

1.3.0 *Any required attributes of the data deck should be as similar as possible for all programs in the library.*

One of the primary virtues of library routines is their flexibility with respect to the layout of input data; nevertheless, some general layout restrictions are inevitable. For example, subject-by-subject (rather than variable-by-variable) data input is usually standard.

1.3.1 *The standard location of the subject identification field should be the first columns of each data card.*

Further standardization of the subject identification field could, of course, be required. For example, the location of a deck code could be standardized. Since most library programs do not require input of the subject identification field, standardization of the field is probably of most importance to the user in the offline management of his data decks.

1.3.2 *Decks containing data from multiple (i.e., repeated) measurements of multiple variables should be organized so that data fields are grouped by measurements, with the order of variables the same within each measurement group.*

As an example of how a data deck might be formatted in accordance with this recommendation, consider a card layout resulting from three administrations of four tests (where A refers to administration and T refers to test):

Column	*Entry*
1-10	Identification field
11-12	A1T1 score
13-14	A1T2 score
15-16	A1T3 score
17-18	A1T4 score
19-20	A2T1 score
21-22	A2T2 score
23-24	A2T3 score
25-26	A2T4 score
27-28	A3T1 score
29-30	A3T2 score

31-32	A3T3
33-34	A3T4
35-80	Blank

OUTPUT STANDARDIZATION

2.0.0 *The organization of output should be as similar as possible for all programs in the library.*

Even after he has acquired the skills required to set up library programs, the user often finds himself unable to interpret the output from a given library routine. One possible source of his difficulty is failure to master the statistical techniques associated with the program, and there is little the computer center can do to alleviate problems arising from such a handicap. On the other hand, the computer center *can* strive to maximize the transfer of "output interpretation techniques" from one library program to another.

2.1.0 *Certain items should appear in the same order in the output of every program in the library.*

It is recommended that the following items appear in the output of every program in the order listed:

1. Program name
2. Program version date
3. Name of computer center
4. Location of computer center
5. User-supplied titles
6. Parameter card entries and meaning of entries
7. User-supplied format cards
8. Other echo checks as appropriate (e.g., missing-data signal card)
9. Results computed by program
10. Confirmation of normal termination

2.2.0 *Results computed by the library program should be clearly labeled in widely known terms instead of cryptic abbreviations.*

Suppose, for example, a matrix of statistics to be printed is known as the "matrix of partial inner multidimensional products" among individuals knowledgeable about the statistical technique. Suppose also that in the definitive article about the technique, the same matrix is referred to as the Q matrix. The programmer should resist the temptation to label the matrix simply "Q MATRIX." Instead, the full name of the matrix should appear in the program's output.

DOCUMENTATION STANDARDIZATION

3.0.0 *Descriptions of library programs should be designed to minimize the user's need for consultative support.*

This view of the proper function of documentation has been discussed in detail in Chapter 3.

3.1.0 *The program library user's manual should contain introductory sections devoted to an overview of the library.*

As described in Chapter 3, suggested section headings would be the following:

    1. Introduction
    2. Data preparation
    3. System control card preparation
    4. Program control card preparation

The last section on program control card preparation would contain descriptions of those program control cards common to most of the programs in the library.

3.2.0 *Program descriptions should be as similar as possible for all programs in the library.*

Recommended sections of program descriptions are listed below in the order suggested in Chapter 3.

3.2.1 *The write-up of each library program should contain a general description.*

The general description should contain (1) a list of the statistics computed by the program and/or a description of the purpose of the program; (2) a list of the contents of the output of the program; (3) description of significant limitations of the program; and (4) suggestions for estimating running time.

3.2.2 *The write-up of each program should contain a section listing the order of cards in the job deck.*

In this section all program control cards should be listed separately, but system control cards should be shown as groups (i.e., lumped together).

3.2.3 *The write-up of each program should contain a section describing the preparation of program control cards not described in the introductory section of the library user's manual.*

This section should also contain brief comments on the required data card arrangement and appropriate format card specifications.

3.2.4 *The write-up of each program should contain a section listing appropriate references describing the statistical techniques implemented by the program.*

The references listed should be easy to obtain and readable.

3.2.5 *The write-up of each program should contain a section presenting any necessary computational notes.*

These notes should clarify any ambiguities concerning the formulas used in the program. In particular, any nonstandard formulas or procedures used should be described.

3.2.6 *The write-up of each program should contain a listing of a sample problem deck and corresponding program output.*

As mentioned in Chapter 3, the problem should exercise most of the program's features.

3.2.7 *The write-up of each program should be dated and contain the name(s) of the person(s) preparing the program and write-up.*

Indication of authorship may appear at either the beginning or end of the write-up. Each page of the write-up should be dated.

3.2.8 *Standard terminology should be used throughout all documentation.*

While use of jargon should be minimized, any use of terms should be consistent.

SOURCE DECK STANDARDIZATION

4.0.0 *The organization of the source deck should be as similar as possible for all programs in the library.*

The source deck is probably more difficult to subject to standardization than input, output, or documentation; however, the resulting improvement in exportability of the routine and communication among computer center programmers more than justifies the effort.

4.1.0 *The first mainline program associated with each library routine should begin with a set of comment cards containing vital program information.*

Items recommended for inclusion among these comment cards appear below in the order recommended.

4.1.1 *The name and title by which the program is known should be included.*

4.1.2 *The name and location of the computer center should be included.*

4.1.3 *The name(s) of program author(s) and version date should be included.*

4.1.4 *If the program has been adapted from a program developed elsewhere, an appropriate note should be included.*

4.1.5 *The card order for the job deck with brief notes on specific program control card preparation should be included.*

This section of comment cards functions as a terse program write-up.

4.1.6 *A list of required subprograms and/or associated mainline programs should be included.*

By "associated mainline program" is meant any routine to which chaining occurs.

4.1.7 *Definition of major array and variable names should be included.*

The relationship of array dimensions to problem parameters should be indicated in this set of comment cards.

4.1.8 *Mention should be made of any special devices (e.g., magnetic tape) or operating procedures required.*

4.1.9 *The model of the computer and version of the operating system under which the program was tested should be included.*

If appropriate, core memory requirements should appear here also.

4.2.0 *Array and variable names should be as similar as possible in all programs in the library.*

The following are a few illustrative names:

Name	Meaning
KARDS	Card reader device number
KF	Variable format specifications (array)
KPNCH	Card punch device number
LINES	Line printer device number
NCS	Number of cards per subject
NFC	Number of format cards
NG	Number of groups
NS	Number of subjects
NT	Number of trials
NTAPE	Magnetic tape device number
NV	Number of variables
R	Correlation matrix (array)
SUMX	ΣX (array)
X	Data matrix or one subject's set of scores (array)

Other array and variable names would, of course, be suggested by the notational scheme used in connection with a specific statistical procedure.

4.3.0 *Comment cards should be used extensively throughout the source deck.*

A considerable amount of grief can be avoided if the programmer thoughtfully documents his work as he creates the program.

4.4.0 *Each subprogram or mainline used only in connection with a single library routine should contain comment cards to that effect.*

These comment cards should contain the name of the library routine, version date, name and location of the computer center, and programmer name(s).

4.5.0 *Source code should include spaces for increased readability.*

The packing of source statements practiced by some programmers, while admittedly facilitating keypunching, exacts a heavy toll in eyestrain from the unfortunate reader.

4.6.0 *Statement numbers should be assigned in ascending numerical sequence with a standard increment.*

Use of an increment of 5 or 10 facilitates later program revisions.

4.7.0 *Source decks should be punched with a deck identification code and sequence number in columns 73 through 80.*

Many computer centers have available a utility program for punching a new deck with an identification code and sequence numbers. In some instances, the same program may also provide the option of renumbering program statements and corresponding program references.

4.8.0 *Only conservative language features should be used in writing library programs.*

The exportability of library programs is greatly enhanced by careful restriction of language features used. For example, the use of 6H MEANS instead of ' MEANS ' or * MEANS* avoids possible program conversion problems. Restriction of array and variable names to five characters is another thoughtful practice. Although it is a considerable sacrifice for the user of a large machine, IBM 1130 users would be particularly delighted to encounter a program in which only arithmetic (i.e., three-branch) IF statements were used, since logical IF statements are not (as of January 1, 1972) available in IBM 1130 FORTRAN.*

Arguing against excessive restriction of language features as a concession to the users of small computing systems is the strong likelihood that a program written for use on a large-scale computing system would require extensive reorganization and rewriting (possibly along the lines suggested in Chapter 4) before it could be adapted for use on a small computer. In such an event, the already extensive source-code modifications could incorporate syntax restrictions without a great increase in programming effort.

_____

*However, in early 1972, the Department of Computer Science of Eastern Michigan University began distributing a new FORTRAN compiler for the IBM 1130. Among other language improvements, the EMU compiler includes logical IF statements.

SUMMARY

The general standardization guidelines presented in this chapter are intended to provide personnel responsible for library program development with at least a starting point.

Program authors at a particular installation may wish to modify these recommendations or add more detailed conventions along the lines suggested by Perry and Sommerfeld (1970).

8 Case Studies of Illustrative Library Programs

Up to this point our examination of library program creation has tended to focus upon bits and pieces of technique and conventions without an opportunity to consider a complete library program of practical proportions. In this, the final chapter of this volume, we look at programs incorporating a number of the suggestions made in previous chapters. Again, as throughout the entire book, the routines presented in this chapter are intended to serve as models to be adapted to the requirements of a particular environment.

PEARSON PRODUCT-MOMENT CORRELATION

The documentation for program C01 was presented in Chapter 3; now we shall consider the actual program described in the write-up. Correlational techniques constitute an important part of any statistical program library, and a basic routine for the calculation of a matrix of Pearson product-moment correlations is likely to see heavy use.

Program C01 was written for an IBM 1130 with 8192 words of core memory. This severely restricted core memory makes exploitation of the disk-operating system imperative, and program C01 reflects this orientation. An inspection of Figure 8-1 reveals that program C01 is actually a set of four mainline programs, C01, C01A, C01B, and C01C, with associated subroutines. Two levels of program segmentation are employed: mainline-to-mainline chaining (the IBM

135

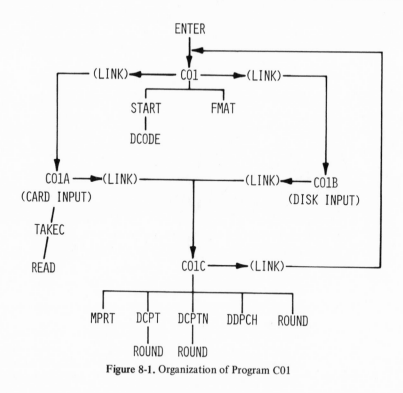

Figure 8-1. Organization of Program C01

1130 LINK feature) and disk overlays involving subprograms (the IBM 1130 LOCAL feature). The source listings in this chapter include all system control cards so that the implementation of these features may be studied in detail.

Statistical Procedure

BASIC FORMULAS

If we define ΣX as the sum of a set of X scores, we may define the *mean,* $\overline{X},$ of the same set of scores by the formula

$$\overline{X} = \frac{\Sigma X}{N} \tag{8-1}$$

where N is the number of scores in the set. Using the same notational scheme, the deviation of any X score from the mean of the X scores is

$$x = X - \overline{X} \tag{8-2}$$

where x is known as a *deviation score,* and the formula

$$\sigma_x{}^2 = \frac{\Sigma x^2}{N} \tag{8-3}$$

may be used to define the *variance* of a set of scores. We shall refer to σ_x, the square root of the variance defined by Formula 8-3, as the *sigma* of a set of scores.

The *covariance* between X and Y is given by the formula

$$\sigma_{xy} = \frac{\Sigma xy}{N} \tag{8-4}$$

where σ_{xy} is the covariance between X and Y, Σxy is the sum of products of deviation scores for X and Y, and N is the number of subjects.

The basic formula for the Pearson product-moment coefficient of correlation (Guilford, 1965) is usually given as

$$r_{xy} = \frac{\Sigma xy}{N\sigma_x\sigma_y} \tag{8-5}$$

where r_{xy} is the correlation between X and Y, σ_y is the sigma of the Y scores, and Σxy, N, and σ_x are defined as above. An equivalent formula often used in place of Formula 8-5 is

$$r_{xy} = \frac{N\Sigma XY - (\Sigma X)(\Sigma Y)}{([N\Sigma X^2 - (\Sigma X)^2][N\Sigma Y^2 - (\Sigma Y)^2])^{1/2}} \tag{8-5a}$$

where X and Y are raw scores. This formula permits calculation of r_{xy} without first having to calculate the means of X and Y.

COMPUTATIONAL FORMULAS AND ACCURACY OF CALCULATIONS

In terms of computer processing, Formula 8-5a would at first appear preferable to Formula 8-5, since two passes over the data would be required for calculation of r_{xy} by Formula 8-5. During input of the data, ΣX and ΣY would be accumulated, and the means of X and Y would be calculated. The raw data would then be re-read (from magnetic tape or disk) and Σxy, Σx^2, and Σy^2 would be accumulated. At this point, Formula 8-5 could be used to calculate r_{xy}. The attractiveness of Formula 8-5a arises from the fact that one pass over the data would suffice for accumulation of ΣX, ΣY, ΣX^2, ΣY^2, and ΣXY, terms required in Formula 8-5a.

The first-blush appeal of Formula 8-5a over Formula 8-5 is moderated, however, by an examination of computational problems associated with Formula 8-5a. Because many computers (e.g., Hewlett-Packard 2000 Series systems, IBM 1130, IBM 360) provide only six-to-seven-digit accuracy in standard precision real arithmetic, the accumulation of such terms as ΣX^2 and ΣXY may involve troublesome inaccuracies. Consider, for example, 1000 scores with a mean of 1000 and a sigma of 50. Another formula for the variance of a set of scores is

$$\sigma_x{}^2 = \frac{\Sigma X^2}{N - \overline{X}^2} \tag{8-6}$$

Substituting the values given for our set of scores, we can write

$$50^2 = \frac{\Sigma X^2}{1000 - 1000^2}$$

which, after suitable manipulations, gives us ΣX^2 = 1,002,500,000. Since the accuracy of FORTRAN is likely to be only six to seven significant digits, any information that might be represented in the last three or four digits of ΣX^2 would be lost. The sigma of 50 for our hypothetical set of scores implies that there is important variation in the last two digits of each score (since, if drawn from a normal distribution, more than 99 percent of our scores would be expected to lie between 850 and 1150), so we are in a vulnerable position with respect to the effects of computational inaccuracies. What happens if we convert to deviation scores and accumulate Σx^2 instead of ΣX^2?

Because $x = X - \overline{X}$, our scores ranging primarily from 850 and 1150 would be converted to deviation scores ranging primarily from -150 to +150. Since $\sigma_x{}^2 = \Sigma x^2/N$, we may write the equation

$$50^2 = \frac{\Sigma x^2}{1000} \quad \text{or} \quad 2500 = \frac{\Sigma x^2}{1000}$$

giving us Σx^2 = 2,500,000. This quantity is considerably more manageable in terms of FORTRAN accuracy, and we lose, at most, one digit. In short, computational considerations would seem to dictate the use of Formula 8-5 and sums of deviation scores instead of Formula 8-5a and sums of raw scores.

COMPUTATIONAL COMPROMISE

It so happens that there exists a middle ground between Formulas 8-5 and 8-5a. Starting again with our set of 1000 scores, let us subtract a constant C from each score in the set. It can be shown that

$$\overline{X}' = \overline{X} - C$$

where $\overline{X}'$ is the new mean of our set of scores, and that the sigma of our set of scores would be unchanged. If we should happen to select C such that $C = \overline{X}$, we would simply convert our set of raw scores to deviation scores.

If we were to carry our experimentation with constants a step further and select constants C and D to subtract from each X and Y score, respectively, we would find r_{xy} unchanged. We may further observe that if $C = \overline{X}$ and $D = \overline{Y}$, then $\Sigma XY = \Sigma xy$, $\Sigma X^2 = \Sigma x^2$, and $\Sigma Y^2 = \Sigma y^2$.

With these relationships in mind, let us choose as C and D the actual X and

Y scores for the first subject in our sample. If our first subject's scores are reasonably close to the means of their respective distributions, we shall have approached the computational benefits of a deviation-score formula (8-5) with the one-pass convenience of a raw score formula (8-5a).

REVISION OF FORMULAS AND EXTENSION TO MORE THAN TWO VARIABLES

In order to make the transition from formulas in terms of X and Y to a FORTRAN program in which all possible pairings of NV variables are considered, it is helpful to replace the symbols X and Y with X_i and X_j. Using the symbol C_i to refer to the first subject's X_i, we can write the formulas used in program C01:

$$X_i = \frac{\Sigma(X_i - C_i)}{N} + C_i \tag{8-7}$$

$$\sigma_{x_i}^2 = \frac{\Sigma(X_i - C_i)^2}{N} - \left[\frac{\Sigma(X_i - C_i)}{N}\right]^2 \tag{8-8}$$

$$\sigma_{x_i x_j} = \frac{\Sigma(X_i - C_i)(X_j - C_j)}{N} - \left[\frac{\Sigma(X_i - C_i)}{N} \frac{\Sigma(X_j - C_j)}{N}\right] \tag{8-9}$$

$$r_{x_i x_j} = \frac{\sigma_{x_i x_j}}{\sigma_{x_i} \sigma_{x_j}} \tag{8-10}$$

C01: The First Phase of Processing

As can be seen in Figure 8-2, most of the code in program C01 is devoted to input of program control cards and error checking. Input of title and parameter cards is accomplished through a call to subroutine START, and subroutine FMAT is used to encode the contents of the format card(s). Both routines are discussed in Chapter 2.

INTRODUCTORY SECTION

Figure 8-3 contains the IBM 1130 system control cards required to invoke the FORTRAN compiler and also lists the comment cards used to document vital program information. The // FOR card fetches the FORTRAN compiler from disk; the *ONE WORD INTEGERS card causes integer variable and arrays to occupy one-half of the core memory they would otherwise require; *LIST

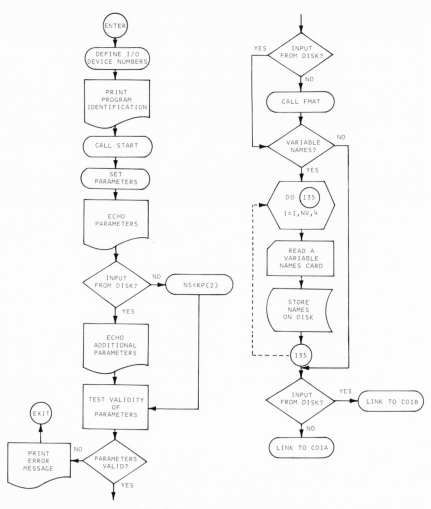

Figure 8-2. Flow Chart for C01

SOURCE PROGRAM causes source statements to be listed on the line printer; and the *IOCS card lists I/O devices to be supported by the program.

The comment cards follow the scheme suggested in Chapter 7. The presence of these cards in the source deck provides a convenient reference during program testing.

The description of the card order for the job deck is intended to provide enough detail for a programmer to set up a card-oriented test problem without consulting the write-up in the user's manual. The only aspect of program operation not covered is the implementation of disk input. The programmer interested in this feature is assumed to have access to the special write-up presented in Chapter 3.

```
// *
// *    C01 SYSTEM CONTROL CARDS AND COMMENT CARDS
// *
// FOR     PROGRAM C01
*ONE WCRD INTEGERS
*LIST SOURCE PROGRAM
*IOCS(CARD,1403PRINTER,DISK)
C     PROGRAM C01, PEARSCN PRODUCT-MOMENT CORRELATION
C     GECRGE PEABODY COLLEGE CCMPUTER CENTER, NASHVILLE, TENNESSEE
C     JAMES H. HOGGE, 1/1/72
C
C     CARD ORDER FOR JOB DECK...
C
C          SYSTEM CONTRCL CARDS
C          TITLE CARD(S) (OPTIONAL)
C               EACH TITLE CARD MUST HAVE AN ALPHABETIC CHARACTER IN
C               CCLUMN 1.
C          PARAMETER CARD...
C               CCL. 1-4      NUMBER OF VARIABLES (MAX = 50)
C               COL. 5-8      NUMBER OF SUBJECTS (MAX = 1000 FOR DISK
C                             INPUT)
C               CCL. 9-12     NAMES CARD(S) SIGNAL...
C                             1 = VARIABLE NAMES TO BE READ
C                             0 = NO NAMES SUPPLIED
C               CCL. 13-16    CUTPUT OPTIONS FCR CCVARIANCE MATRIX...
C                             0 = NO OUTPUT
C                             1 = PRINT ONLY
C                             2 = PUNCH ONLY
C                             3 = PRINT AND PUNCH
C               CCL. 17-20    CUTPUT OPTIONS FCR CCRRELATICN MATRIX...
C                             (SAME AS ABOVE)
C               CCL. 21-24    DISK INPUT OPTION...
C                             1 = DISK INPUT
C                             0 = CARD INPUT
C                             FCR DISK INPUT, SEE SPECIAL WRITE-UP.
C               CCL. 25-80    BLANK
C          FORMAT CARC(S)...
C               SPECIFY CNLY VARIABLE FIELDS IN F-NOTATION.
C               FCRMAT MAY BE CCNTINUED ON SECOND CARD BY PUNCHING AN
C               ASTERISK (*) IN COLUMN 80 OF FIRST FORMAT CARD.
C          VARIABLE NAMES CARC(S) (CPTIONAL)...
C               LEFT-JUSTIFIED VARIABLE NAMES APPEAR ON ONE OR MORE
C               CARDS IN 20-CCLUMN FIELDS IN SAME ORDER AS ORDER CF
C               INPUT.
C          DATA CARDS
C          BLANK CARDS FCR PUNCHEC CUTPUT (IF SPECIFIED)...
C               INCLUDE V*(V+1)/14 (WHERE V = NUMBER OF VARIABLES) BLANK
C               CARDS FOR EACH MATRIX TO BE PUNCHED.
C          BLANK ENC-CF-JCB CARC (STCPS PROGRAM)...
C               TC ANALYZE MCRE THAN ONE DATA SET IN ONE RUN, INSERT
C               TITLE CARC(S), ETC., FOR NEXT SET INSTEAD OF BLANK
C               CARD.
C          SYSTEM CONTROL CARDS
C
C     MAINLINES AND SUBPROGRAMS REQUIRED ARE C01A, C01B, C01C, DCODE,
C     DCPT, DCPTN, CCPCH, FMAT, MPRT, ROUND, START, AND TAKEC.
C     SUBRCUTINE REAC (FRCM THE IBM COMMERCIAL SUBRCUTINE PACKAGE)
C     IS ALSC REQUIRED.
C
C     CEFINITICNS OF MAJCR VARIABLE AND ARRAY NAMES...
C
C     VARIABLE          MEANING
C
C     ICPTA             OUTPUT OPTION FOR COVARIANCE MATRIX
C     ICPTB             CUTPUT CPTION FOR CORRELATION MATRIX
C     KARDS             CEVICE NUMBER FOR CARD READER
C     KPNCH             DEVICE NUMBER FOR CARD PUNCH
C     LINES             CEVICE NUMBER FOR LINE PRINTER
C     NAMES             VARIABLE NAMES SIGNAL
C     NDISK             DISK INPUT SIGNAL
C     NS                NUMBER OF SUBJECTS
C     NV                NUMBER CF VARIABLES
C
C     ARRAY AND DIMENSION              MEANING
C
C     KF(3*(NV+10))                    FORMAT SPECIFICATIONS
C     KP(80)                           PARAMETERS RETURNED BY START,
C                                      LATER USED FCR TEMPORARY
C                                      STCRAGE OF VARIABLE NAMES.
C     SUB(NV)                          FIRST SUBJECT'S SCCRES
C     SUMX(NV)                         SUMS OF DEVIATIONS ABCUT FIRST
C                                      SUBJECT'S SCORES, LATER MEANS
C     R((NV*NV-NV)/2+NV)               SUMS OF CROSS-PRODUCTS OF
C                                      DEVIATIONS ABOUT FIRST SLBJECT'S
C                                      SCCRES.
```

Figure 8-3. C01 System Control Cards and Comment Cards

```
C         X(NV)                          TEMPORARY STORAGE FOR EACH
C                                             SUBJECT'S SCORES
C
C    DISK FILES...
C
C    FILE              SIZE                         CONTENTS
C
C     1        NS 2*NV-WCRD RECORDS              DATA FCR DISK INPUT
C     2        NV+2 20-WCRD RECORDS                 VARIABLE NAMES
C     3        ENCUGH 320-WORD RECORDS TO HOLD NV REAL ARRAY ELEMENTS
C     4        ENCUGH 320-WCRD RECORDS TO HCLD ARRAY R -- LATER
C              FQLIVALENT SIZE IN 2-WORD RECORDS
C
C    THIS ROUTINE TESTED UNDER VERSION 2, MODIFICATION LEVEL 7, OF THE
C        IBM 1130 CISK OPERATING SYSTEM CN AN 8-K SYSTEM.
C
```

Figure 8-3. (continued)

Included as part of the definitions of array names are the dimensions of the arrays in the form of expressions involving the integer variable names defined in the preceding section. A programmer attempting to convert the program for operation on another system would be able to ascertain that the dimension of SUMX would have to be changed from 50 in the present program to 100 in a revision designed to accommodate twice as many variables. In a similar fashion, the requirements for disk files are described in terms of problem parameters so that program modification is facilitated.

PROGRAM LOGIC

The remaining source statements of program C01 appear in Figure 8-4. As suggested in Chapter 7, comment cards are used freely in an effort to provide a running description of program logic.

Variables and arrays to be transmitted to subsequent program segments have been placed in COMMON. Because the array KP is needed only in program C01, it is declared in a DIMENSION statement.

After subroutine START has been used to read and print any title cards, read and decode the parameter card, and store parameters in the array KP, the variable names to be used to represent problem parameters throughout the program are equated to the appropriate elements of KP. The WRITE statement referencing FORMAT 15 is required because IBM 1130 FORTRAN permits only five statement continuation cards.

The provision for disk input of data for program C01 permits the programmer to transmit a value for NS via COMMON. If, on the other hand, card input is specified, a value for NS is assumed to have been punched in columns 5 through 8 of the parameter card. After subroutine START has processed the parameter card, that same value appears in KP(2). The IF statement before statement 20 is designed to test for the possible specification of disk input and branch to appropriate statements.

The sequence of statements beginning with statement 35 provides tests of the plausibility of user-supplied parameter card entries. The error messages as-

```
C
C      C01 -- BODY OF PROGRAM
C
       DIMENSION KP(80)
       COMMON NS, IOPTA, IOPTB, KARDS, KPNCH, LINES, NAMES, NDISK, NV,
      1 KF(180)
       DEFINE FILE 1 (1000,100,U,N1), 2 (52,20,U,N2), 3 (1,320,U,N3),
      1 4 (8,320,U,N4)
C
C      DEFINE I/O DEVICE NUMBERS.
C
       KARDS = 2
       KPNCH = 2
       LINES = 5
C
C      PRINT PROGRAM IDENTIFICATION.
C
       WRITE (LINES,5)
     5 FORMAT (67H1PROGRAM C01, PEARSON PRODUCT-MOMENT CORRELATION, VERSI
      1ON OF 1/1/72 / 61H GEORGE PEABODY COLLEGE COMPUTER CENTER, NASHVIL
      2LE, TENNESSEE /)
C
C      INPUT TITLE AND PARAMETER CARDS.
C
       CALL START (KP,KARDS,LINES)
C
C      DEFINE PARAMETERS.
C
       NV = KP(1)
       NAMES = KP(3)
       IOPTA = KP(4)
       IOPTB = KP(5)
       NDISK = KP(6)
C
C      ECHO PARAMETERS.
C
       WRITE (LINES,10) (KP(I), I = 1,4)
    10 FORMAT (23H0PARAMETER CARD ENTRIES / 24H COLUMNS   ENTRY   MEANING /
      1 6H0   1-4, I8, 3X, 19HNUMBER OF VARIABLES / 6H    5-8, I8, 3X,
      2 18HNUMBER OF SUBJECTS / 7H    9-12, I7, 3X,
      3 21HVARIABLE NAMES OPTION / 7H   13-16, I7, 3X,
      4 35HCOVARIANCE MATRIX OUTPUT OPTIONS... / 20X, 13H0 = NO OUTPUT /
      5 20X, 14H1 = PRINT ONLY / 20X, 14H2 = PUNCH ONLY)
       WRITE (LINES,15) KP(5)
    15 FORMAT (20X, 19H3 = PRINT AND PUNCH / 7H   17-20, I7, 3X,
      1 36HCORRELATION MATRIX OUTPUT OPTIONS... / 20X,
      2 15H(SAME AS ABOVE))
C
C      CHECK FOR DISK INPUT, SET PARAMETERS, AND ECHO, IF REQUIRED.
C
       IF (NDISK) 20,20,25
    20 NS = KP(2)
       GO TO 35
    25 WRITE (LINES,30) NDISK, NS
    30 FORMAT (7H   21-24, I7, 3X, 10H0DISK INPUT / 20X, 23H(FIRST WORD IN
      1COMMON = I6, 1H))
C
C      TEST VALIDITY OF PARAMETERS.
C
    35 IF (NV) 40,40,50
    40 WRITE (LINES,45)
    45 FORMAT (26H0ZERO VARIABLES SPECIFIED.)
       CALL EXIT
    50 IF (NV - 50) 65,65,55
    55 WRITE (LINES,60)
    60 FORMAT (31H0TOO MANY VARIABLES (MAX = 50).)
       CALL EXIT
    65 IF (NS) 70,80,90
    70 WRITE (LINES,75)
    75 FORMAT (56H0FAULTY NUMBER OF SUBJECTS SPECIFICATION FOR DISK INPUT
      1.)
       CALL EXIT
    80 WRITE (LINES,85)
    85 FORMAT (25H0ZERO SUBJECTS SPECIFIED.)
       CALL EXIT
    90 IF (IOPTA - 3) 105,105,95
    95 WRITE (LINES,100)
   100 FORMAT (30H0INVALID MATRIX OUTPUT OPTION.)
       CALL EXIT
   105 IF (IOPTB - 3) 110,110,95
C
C      INPUT AND ECHO FORMAT CARD(S) IF INPUT IS FROM CARDS.
C
   110 IF (NDISK) 115,115,120
   115 CALL FMAT (KF,KARDS,LINES,60)
C
C      INPUT VARIABLE NAMES, IF OPTED.
```

Figure 8-4. C01—Body of Program

```
C
   120 IF (NAMES) 140,140,125
   125 DO 135 I = 1,NV,4
       READ (KARDS,130) KP
   130 FORMAT (80A1)
   135 WRITE (2'I) KP
C
C       CHECK FOR DISK INPUT, AND TRANSFER CONTROL TO APPROPRIATE SEGMENT.
C
   140 IF (NDISK) 145,145,150
   145 CALL LINK (C01A)
   150 CALL LINK (C01B)
       END
// DUP
*STORECI    WS  UA  C01      1
*LOCAL,START,FMAT
```

Figure 8-4. (continued)

sociated with this section of the program are designed to be as specific as possible.

If data input is to be from cards, the call to subroutine FMAT at statement 115 results in the reading of one or two format cards and encoding of specifications from those cards in the array KF. If disk input has been specified, statement 115 is skipped.

Because variable names are punched four per card, the DO 135 loop is indexed from 1 to NV in increments of 4. This indexing causes the appropriate number of cards to be read and provides the correct succession of record numbers for statement 135, the disk WRITE.

The last IF statement in C01 causes control to be transferred to C01A if input is to be from cards, or to C01B if input is to be from disk.

The system control cards at the end of C01 cause the program to be stored on disk as an executable file. The // DUP card fetches the disk utility program from disk. Then DUP reads the *STORECI and *LOCAL cards. The *STORECI card instructs DUP to move a program named C01 from working storage (WS) to the user area (UA) of the disk, storing it in core image (CI). Stored in this form, C01 can be fetched from disk and "turned on" by a single // XEQ C01 card. The *LOCAL card specifies that subroutines START and FMAT are to overlay one another in an area in core memory.

Program C01A: Input from Cards

The flow chart presented in Figure 8-5 portrays the functions performed by C01A. Program C01A has been kept as simple as possible to minimize core requirements.

SYSTEM CONTROL CARDS

The source deck for C01A appears in Figure 8-6. The system control cards at the beginning of the deck are very similar to those at the beginning of C01; only the *IOCS card differs in terms of control effects (the information punched

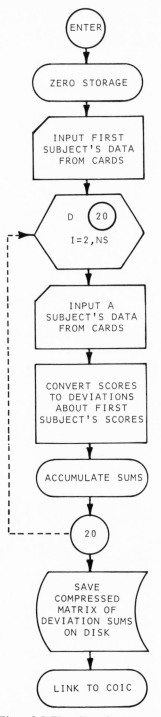

Figure 8-5. Flow Chart for C01A

```
// *
// *  CO1A
// *
// FOR    PROGRAM CO1A (LINK FRCM CO1)
*ONE WCRD INTEGERS
*LIST SOURCE PROGRAM
*IOCS(1403PRINTER,DISK)
C     PROGRAM CO1A, LINK FRCM PROGRAM CO1
C
C     THIS ROUTINE INPUTS DATA FROM CARDS AND ACCUMULATES SUMS.
C
C     GEORGE PEABCCY CCLLEGE CCMPUTER CENTER, NASHVILLE, TENNESSEE
C     JAMES H. HCGGE, 1/1/72
C
      COMMON NS, IOPTA, IOPTB, KARDS, KPNCH, LINES, NAMES, NDISK, NV,
     1 KF(180), SUB(50), SUMX(50), X(50), R(1275)
      CEFINE FILE 1 (1000,1CC,U,N1), 2 (52,20,U,N2), 3 (1,320,U,N3),
     1 4 (8,320,U,N4)
C
C     ZERC STORAGE.
C
      NTOT = (NV * NV - NV) / 2 + NV
      DO 5 I = 1,NTCT
    5 R(I) = 0.0
      CC 10 I = 1,NV
   10 SUMX(I) = 0.0
C
C     INPUT FIRST SUBJECT'S CATA FROM CARDS.
C
      CALL TAKEC (KF,SUB,NV,LINES)
C
C     INPUT DATA FCR REMAINING SLBJECTS, CCNVERTING SCCRES TC DEVIATIONS
C         ABOUT FIRST SUBJECT'S SCCRES AND ACCUMULATING SUMS.
C
      DC 20 I = 2,NS
      NE = 0
C
C     INPUT A SUBJECT'S SCCRES FROM CARDS.
C
      CALL TAKEC (KF,X,NV,LINES)
C
C     CONVERT TO DEVIATICN SCCRES.
C
      CC 15 J = 1,NV
   15 X(J) = X(J) - SUB(J)
C
C     ACCUMULATE SUMS.
C
      CC 20 J = 1,NV
      SUMX(J) = SUMX(J) + X(J)
      CC 20 K = J,NV
      NE = NE + 1
   2C R(NE) = R(NE) + X(J) * X(K)
C
C     SAVE MATRIX CN DISK.
C
      WRITE (4'1) (R(I), I = 1,NTCT)
C
C     TRANSFER CCNTRCL TC NEXT SEGMENT.
C
      CALL LINK (CO1C)
      END
// DUP
*STORECI    WS  UA  CO1A      1
*LOCAL,READ,PRNZ,SDFIO,SFIO
```

Figure 8-6. Program C01A (Link from C01)

beginning in column 11 of the // FOR card is only a comment and is ignored by the FORTRAN compiler). Because subroutine TAKEC is used for card input, and subroutine TAKEC references subroutine READ of the IBM commercial subroutine package (discussed in Chapter 6), CARD does not appear in the *IOCS card specifications.

IMPLEMENTATION OF DATA INPUT

In accordance with Formulas 8-7 through 8-10, data input involves storing the first subject's scores in the array SUB, then using the scores in SUB to convert data for subjects 2 through NS to deviation scores. Sums of deviation scores are accumulated in the array R, which is really a one-dimensional representation of the upper triangle of a symmetric matrix. Subroutine TAKEC, through its reference to subroutine READ, takes advantage of the I/O overlap capabilities of the IBM 1130. Input via subroutine TAKEC is dramatically faster than it would be via subroutine TAKE.

After data input has been completed, the array R is stored in disk file 4, and control is transferred to C01C.

The *LOCAL card for C01A is a departure from previous examples in this volume because it includes the names of system subprograms in addition to programmer-supplied routines. While READ is a subroutine of the IBM commercial subroutine package (discussed in Chapter 6) and is called by the TAKEC subroutine, PRNZ, SDFIO, and SFIO are system subprograms. PRNZ is used in printing on the IBM 1403 printer, and both SDFIO and SFIO are disk I/O routines.

This particular combination of LOCALed routines slows down processing comparatively little, since PRNZ is required only in the printing of an (fatal) error message, and the only disk READ or WRITE statement appears at the end of the program outside a loop.

Program C01B: Input from Disk

The flow chart for C01B (Figure 8-7) is very similar to that for C01A; the programs differ only with respect to the source of data during input.

SYSTEM CONTROL CARDS

As may be seen in Figure 8-8, the *IOCS card for C01B refers only to DISK. Both CARD and 1403 PRINTER have been omitted because neither card input/output nor printing occurs in C01B (C01A references subroutine TAKEC, which, if it detects an input error, prints an error message).

DATA INPUT

Program C01B differs from C01A only with respect to the medium from which data are acquired. Where C01A has a call to subroutine TAKEC, C01B has a disk READ statement. Otherwise, the two routines are identical.

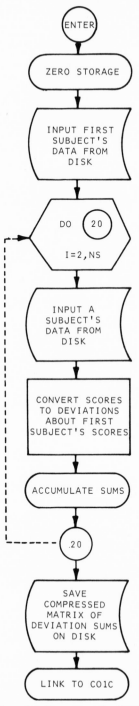

Figure 8-7. Flow Chart for C01B

```
// *
// *    C01B
// *
// FOR     PROGRAM C01B (LINK FROM C01)
*ONE WORD INTEGERS
*LIST SOURCE PROGRAM
*IOCS(DISK)
C       PROGRAM C01B, LINK FROM PROGRAM C01
C
C       THIS ROUTINE INPUTS DATA FROM DISK AND ACCUMULATES SUMS.
C
C       GEORGE PEABODY COLLEGE COMPUTER CENTER, NASHVILLE, TENNESSEE
C       JAMES H. HOGGE, 1/1/72
C
        COMMON NS, IOPTA, IOPTB, KARDS, KPNCH, LINES, NAMES, NDISK, NV,
      1 KF(180), SUB(50), SUMX(50), X(50), R(1275)
        DEFINE FILE 1 (1000,100,U,N1), 2 (52,20,U,N2), 3 (1,320,U,N3),
      1 4 (8,320,L,N4)
C
C       ZERO STORAGE.
C
        NTOT = (NV * NV - NV) / 2 + NV
        DO 5 I = 1,NTOT
      5 R(I) = 0.0
        DO 10 I = 1,NV
     10 SUMX(I) = 0.0
C
C       READ FIRST SUBJECT'S DATA FROM DISK.
C
        READ (1'1) (SUB(I), I = 1,NV)
C
C       INPUT DATA FOR REMAINING SUBJECTS, CONVERTING SCORES TO DEVIATIONS
C           ABOUT FIRST SUBJECT'S SCORES AND ACCUMULATING SUMS.
C
        DO 20 I = 2,NS
        NE = 0
C
C       READ A SUBJECT'S SCORES FROM DISK.
C
        READ (1'I) (X(J), J = 1,NV)
C
C       CONVERT TO DEVIATION SCORES.
C
        DO 15 J = 1,NV
     15 X(J) = X(J) - SUB(J)
C
C       ACCUMULATE SUMS.
C
        DO 20 J = 1,NV
        SUMX(J) = SUMX(J) + X(J)
        DO 20 K = J,NV
        NE = NE + 1
     20 R(NE) = R(NE) + X(J) * X(K)
C
C       SAVE MATRIX ON DISK.
C
        WRITE (4'1) (R(I), I = 1,NTOT)
C
C       TRANSFER CONTROL TO NEXT SEGMENT.
C
        CALL LINK (C01C)
        END
// DUP
*STORECI    WS  UA  C01B
```

Figure 8-8. Program C01B (Link from C01)

Program C01C: Computation of Statistics and Output

Figure 8-9 presents a highly simplified flow chart for C01C. Although the main flow of program logic is depicted, details have been excluded. Missing, for example, is a clear indication that C01C deals with a disk-resident matrix. Because C01C is relatively lengthy, we shall deal with it in two sections.

Figure 8-10 includes source statements corresponding to blocks up to the first decision block ("print covariances?") in the flow chart. Because file number 4 contains the array R (defined in either C01A or C01B) as a disk-resident

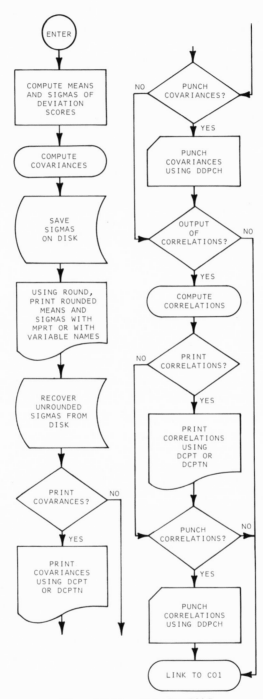

Figure 8-9. Flow Chart For C01C

```
// *
// *   C01C -- FIRST PART
// *
// FOR      PROGRAM C01C (LINK FROM C01A OR C01B)
*ONE WORD INTEGERS
*LIST SOURCE PROGRAM
*IOCS(CARD,1403PRINTER,DISK)
C        PROGRAM C01C, LINK FROM C01A OR C01B (USED WITH C01)
C
C        THIS ROUTINE COMPUTES AND OUTPUTS MEANS, SIGMAS, COVARIANCES,
C            AND/OR CORRELATIONS.
C
C        GEORGE PEABODY COLLEGE COMPUTER CENTER, NASHVILLE, TENNESSEE
C        JAMES H. HOGGE, 1/1/72
C
         DIMENSION IDA(5), IDB(5)
         COMMON NS, IOPTA, IOPTB, KARDS, KPNCH, LINES, NAMES, NDISK, NV,
        1 KF(180), SUB(50), SUMX(50), X(50)
C
C        SET UP IDENTIFICATION FIELDS FOR PUNCHED OUTPUT.
C
         DATA IDA, IDB /'C',' ','M','A','T','R',' ','M','A','T'/
C
         DEFINE FILE 1 (1000,100,U,N1), 2 (52,20,U,N2), 3 (1,320,U,N3),
        1 4 (1280,2,U,N4)
C
C        COMPUTE MEANS AND SIGMAS OF DEVIATION SCORES.
C
         SN = NS
         NE = 1
         DO 5 I = 1,NV
         SUMX(I) = SUMX(I) / SN
C
C        READ SUM OF PRODUCTS FROM DISK.
C
         READ (4'NE) R
C
         X(I) = SQRT(R / SN - SUMX(I) * SUMX(I))
       5 NE = NE + NV - I + 1
C
C        COMPUTE COVARIANCES.
C
         NE = 0
         DO 10 I = 1,NV
         DO 10 J = I,NV
         NE = NE + 1
C
C        READ SUM OF PRODUCTS FROM DISK.
C
         READ (4'NE) R
C
         R = R / SN - SUMX(I) * SUMX(J)
C
C        STORE COVARIANCE ON DISK.
C
      10 WRITE (4'NE) R
C
C        SAVE SIGMAS ON DISK.
C
         WRITE (3'1) (X(I), I = 1,NV)
C
C        PRINT ROUNDED MEANS AND SIGMAS WITH OR WITHOUT VARIABLE NAMES.
C
         IF (NAMES) 15,15,35
C
C        ROUND BEFORE PRINTING, CORRECTING MEANS FOR SUBTRACTION OF FIRST
C            SUBJECT'S DATA.
C
      15 DO 20 I = 1,NV
         CALL ROUND (SUMX(I)+SUB(I),3,SUMX(I))
      20 CALL ROUND (X(I),3,X(I))
C
C        PRINT WITHOUT VARIABLE NAMES.
C
         WRITE (LINES,25)
      25 FORMAT (6HMEANS/)
         CALL MPRT (SUMX,NV,1,LINES,50)
         WRITE (LINES,30)
      30 FORMAT (7HOSIGMAS/)
         CALL MPRT (X,NV,1,LINES,50)
         GO TO 55
C
C        ROUND IN PRINT LOOP, CORRECTING MEANS.
C
      35 WRITE (LINES,40)
      40 FORMAT (55HOVARIABLE        MEAN        SIGMA        DESCRIPTION
        1 /)
         DO 45 I = 1,NV
         CALL ROUND (SUMX(I)+SUB(I),3,SUMX(I))
```

Figure 8-10. Program C01C (Link from C01A or C01B), First Part

```
      CALL ROUND (X(I),3,X(I))
C
C     READ VARIABLE NAME FROM DISK.
C
      READ (2'I) (KF(J), J = 1,20)
C
C     PRINT WITH VARIABLE NAME.
C
   45 WRITE (LINES,50) I, SUMX(I), X(I), (KF(J), J = 1,20)
   50 FORMAT (I5, F17.3, F14.3, 8X, 20A1)
C
C     READ UNROUNDED SIGMAS FROM DISK.
C
   55 READ (3'I) (X(I), I = 1,NV)
```

Figure 8-10. (continued)

compressed symmetric matrix, the array R no longer appears in the COMMON statement. Instead, R simply appears as a nonsubscripted variable.

COMPUTATION OF MEANS, SIGMAS, AND COVARIANCES

In the DO 5 loop, the elements of SUMX are converted to the means of deviation scores and the array X is used for storage of sigmas of deviation scores (which, as mentioned earlier, are the same as sigmas of raw scores). It is important to remember that the term *deviation score*, as used in this context, really refers to deviations about the first subject's scores rather than to the means of the NV variables. The covariances computed in the DO 10 loops are equivalent to the same statistic based on raw scores, so no later conversion is necessary; however, the means must be "corrected" according to Formula 8-7 before printing.

OUTPUT OF MEANS AND SIGMAS

Although not shown in Figure 8-9, two separate sequences of statements for printing with and without variable names appear in C01C. In the sequence for output without variable names, subroutine ROUND is called to accomplish rounding in the DO 20 loop. The IBM 1130 FORTRAN does not round its output; hence, a FORTRAN routine (listed in Appendix A) for rounding has been prepared and may be referenced by a statement of the form

CALL ROUND (X,N,Y)

where   X = real variable or expression to be rounded

N = integer constant, variable, or expression and is the number of decimal places to which to round

Y = real variable, which is returned containing the rounded value

Because unrounded values for the sigmas are required for later conversion of covariances to correlations, the elements of X are saved in disk file 3 prior to rounding.

Subroutine MPRT prints array elements by the specification F9.3; hence, the calls to subroutine ROUND in the DO 20 loop feature a value of 3 for the argument N. The first call to subroutine ROUND results in a rounded value for $\overline{X}_i$ as specified in Formula 8-7; statement 10 yields rounded sigmas.

Printing with variable names occurs in the DO 45 loop. In each successive cycle of the loop a variable name is read from disk file 2, using the array KF for temporary storage.

Finally, the unrounded sigmas are recovered from disk file 3 at statement 55.

The source statements in Figure 8-11 correspond to the blocks of the flow chart, beginning with the "print covariances?" decision block.

OUTPUT OF COVARIANCES

Because both subroutine DCPT and subroutine DCPTN call subroutine ROUND, no rounding of the disk-resident covariance matrix is done in C01C. Punching of the covariance matrix is accomplished using subroutine DDPCH, which has the calling sequence

CALL DDPCH (KFILE,N,KPNCH,MATID)

KFILE = number of the disk file in which a compressed symmetric matrix is stored as a series of two-word records

N = number of rows and columns in the matrix

KPNCH = device number for the card punch

MATID = five-element, one-dimensional integer array containing an identification field to be punched on each card

The identification field is stored one character per element of MATID, and, as in the case of C01C, may be defined in a DATA statement. An example of punched output from subroutine DDPCH appears in Figure 3-2. Written for an IBM 1130 with an IBM 1442 card READ/PUNCH unit, subroutine DDPCH inspects each card to make sure it is completely blank before punching. A listing of subroutine DDPCH appears in Appendix A.

COMPUTATION AND OUTPUT OF CORRELATIONS

If either printed or punched output of the correlation matrix is required, correlations are computed in the DO 125 loop. Included in the computational sequence is a test of $\sigma_{x_i} * \sigma_{x_j}$ for near-zero (resulting in $r_{ij} = 0.0$) and definition of r_{ii} as 1.0. Output of correlations is accomplished in a sequence of statements paralleling the sequence for output of covariances.

Since stacking of data sets is permitted, statement 160 returns control to C01, which looks for a possible additional analysis.

```
C
C     C01C -- SECOND PART
C
C
C     PRINT COVARIANCES, IF REQUIRED.
C
      IF (IOPTA - 1) 95,65,60
   60 IF (IOPTA - 3) 90,65,95
C
C     PRINT TITLE.
C
   65 WRITE (LINES,70)
   70 FORMAT (/12H0COVARIANCES)
      IF (NAMES) 75,80,80
C
C     PRINT WITHOUT VARIABLE NAMES.
C
   75 CALL DCPT (4,NV,LINES)
      GO TO 85
C
C     PRINT WITH VARIABLE NAMES.
C
   80 CALL DCPTN (4,NV,LINES,2)
C
C     PUNCH COVARIANCES, IF REQUIRED.
C
   85 IF (IOPTA - 3) 95,90,95
C
C     PUNCH.
C
   90 CALL DDPCH (4,NV,KPNCH,IDA)
C
C     COMPUTE CORRELATIONS, IF REQUIRED.
C
   95 IF (IOPTB) 160,160,100
  100 NE = 0
      DO 125 I = 1,NV
      DO 125 J = I,NV
      NE = NE + 1
C
C     CHECK FOR DIAGONAL ELEMENT.
C
      IF (I - J) 105,120,105
C
C     CHECK FOR PRODUCT OF SIGMAS NEAR ZERO.
C
  105 PROD = X(I) * X(J)
      IF (PROD - 0.00001) 115,115,110
C
C     READ COVARIANCE FROM DISK.
C
  110 READ (4'NE) R
C
C     CONVERT COVARIANCE TO CORRELATION.
C
      R = R / PROD
      GO TO 125
C
C     SET UNDEFINED CORRELATION EQUAL TO ZERO.
C
  115 R = 0.0
      GO TO 125
C
C     SET DIAGONAL ELEMENT OF CORRELATION MATRIX EQUAL TO ONE.
C
  120 R = 1.0
C
C     STORE CORRELATION ON DISK.
C
  125 WRITE (4'NE) R
C
C     PRINT CORRELATIONS, IF REQUIRED.
C
      IF (IOPTB - 2) 130,160,130
C
C     PRINT TITLE.
C
  130 WRITE (LINES,135)
  135 FORMAT (/13H0CORRELATIONS)
C
C     CHECK FOR VARIABLE NAMES, AND BRANCH ACCORDINGLY.
C
      IF (NAMES) 140,140,145
C
C     PRINT WITHOUT VARIABLE NAMES.
C
```

Figure 8-11. Second Part of Program C01C

```
  140 CALL DCPT (4,NV,LINES)
      GO TO 150
C
C      PRINT WITH VARIABLE NAMES.
C
  145 CALL DCPTN (4,NV,LINES,2)
C
C      PUNCH CORRELATIONS, IF REQUIRED.
C
  150 IF (IOPTB - 3) 160,155,160
C
C      PUNCH.
C
  155 CALL DCPCH (4,NV,KPNCH,ICB)
C
C      TRANSFER CONTROL TO FIRST SEGMENT TO LOOK FOR AN ADDITIONAL
C          DATA SET.
C
  160 CALL LINK (C01)
C
      END
// DUP
*STORECI    WS  UA  C01C
```

Figure 8-11. (continued)

FREQUENCY TABULATION FOR SINGLE-COLUMN VARIABLES

Program D03, also documented in Chapter 3, represents a relatively simple statistical library program. Frequencies and percentages for single-digit variables (such as items in a survey) are in surprisingly high demand; furthermore, the statistical sophistication of a sizable number of program library users is such that the level of "analysis" performed by program D03 is well matched to their training.

Like program C01, program D03 was written for an IBM with 8192 words of core memory. Because it is less complex than program C01, program D03 involves less segmentation. As shown in Figure 8-12, only two mainlines, D03 and

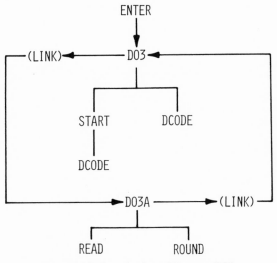

Figure 8-12. Organization of Program D03

D03A, comprise the first level of segmentation of program D03. At the second level of segmentation, subroutines READ and ROUND are LOCALed in the core load for D03A. Subroutines START and DCODE are not LOCALed in the core load for D03 because both mainline D03 and subroutine START call subroutine DCODE.

Statistical Procedure

The means and sigmas computed and printed by program D03 are based on Formulas 8-1 and 8-5a, respectively. It should be noted, however, that only nonzero numerical codes are used in the calculation of the mean and sigma for each variable.

Because only single-column variables are accepted by the program, ΣX^2 for a given variable cannot exceed the product of 9999 (the maximum number of subjects) and 9 (the largest score that can be punched in a single column), or 89,991. This value is within the six- to seven-digit accuracy of standard precision in IBM 1130 FORTRAN, so the extraordinary computational measures taken in program C01 need not be applied here.

Program D03: Control Card Input

The flow chart presented in Figure 8-13 shows that the mainline D03 parallels the mainline C01 in its primary function. The major difference in the control card schemes of the two programs is D03 use of variable location cards rather than the format cards used by C01. Accordingly, D03 does not reference subroutine FMAT.

INTRODUCTORY SECTION

The system control cards and comment cards in Figure 8-14 also demonstrate the similarities in the input characteristics of programs D03 and C01. Because the control card scheme is kept as similar as possible from one PSL routine to the next, many of the comment cards are simply duplicated for insertion in various programs of the PSL.

Where possible, the same variable and array names have also been used in both programs. This practice benefits computer center personnel engaged in library program development by improving communication among programmers and making it easier for a programmer to return to a given program after having set it aside for an extended period of time.

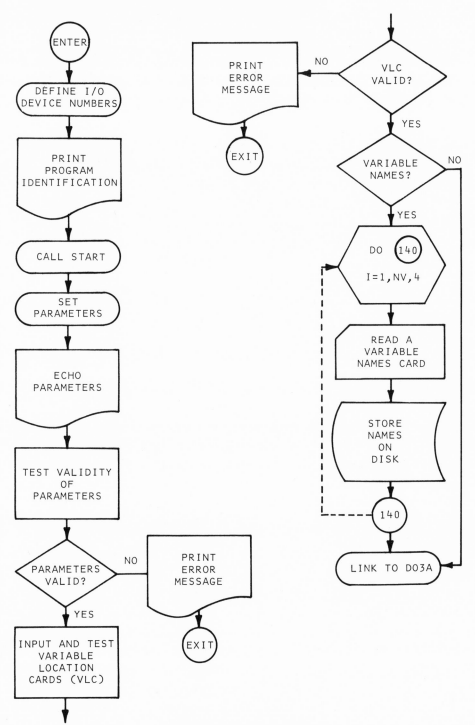

Figure 8-13. Flow Chart for D03

```
// *
// *    D03 -- SYSTEM CONTROL CARDS AND COMMENT CARDS
// *
// FOR      PROGRAM D03
*ONE WORD INTEGERS
*LIST SOURCE PROGRAM
*IOCS(CARD,1403PRINTER,DISK)
C     PROGRAM D03, FREQUENCY TABULATION FOR SINGLE-COLUMN VARIABLES
C     GEORGE PEABODY COLLEGE COMPUTER CENTER, NASHVILLE, TENNESSEE
C     JAMES H. HOGGE, 1/1/72
C
C     CARD ORDER FOR JOB DECK...
C
C         SYSTEM CONTROL CARDS
C         TITLE CARD(S) (OPTIONAL)
C             EACH TITLE CARD MUST HAVE AN ALPHABETIC CHARACTER IN
C             COLUMN 1.
C         PARAMETER CARD...
C             COL. 1-4      NUMBER OF VARIABLES (MAX = 100)
C             COL. 5-8      NUMBER OF SUBJECTS
C             COL. 9-12     NAMES CARD(S) SIGNAL...
C             COL. 13-16    NUMBER OF CARDS PER SUBJECT
C             COL. 17-80    BLANK
C         VARIABLE LOCATION CARD(S)...
C             A ''DUMMY SUBJECT CARD SET'' IS PUNCHED WITH ONES IN
C             THOSE COLUMNS FROM WHICH SCORES ARE TO BE READ, AND ALL
C             OTHER COLUMNS ARE LEFT BLANK.
C         VARIABLE NAMES CARD(S) (OPTIONAL)...
C             LEFT-JUSTIFIED VARIABLE NAMES APPEAR ON ONE OR MORE
C             CARDS IN 20-COLUMN FIELDS IN SAME ORDER AS ORDER OF
C             INPUT.
C         DATA CARDS
C         BLANK END-OF-JOB CARD (STOPS PROGRAM)...
C             TO ANALYZE MORE THAN ONE DATA SET IN ONE RUN, INSERT
C             TITLE CARD(S), ETC., FOR NEXT SET INSTEAD OF BLANK
C             CARD.
C         SYSTEM CONTROL CARDS
C
C     MAINLINES AND SUBPROGRAMS REQUIRED ARE D03B, DCODE, ROUND, AND
C     START.  SUBROUTINE READ (FROM THE IBM COMMERCIAL SUBROUTINE
C     PACKAGE) IS ALSO REQUIRED.
C
C     DEFINITIONS OF MAJOR VARIABLE AND ARRAY NAMES...
C
C     VARIABLE         MEANING
C
C     KARDS            DEVICE NUMBER FOR CARD READER
C     LINES            DEVICE NUMBER FOR LINE PRINTER
C     NAMES            VARIABLE NAMES SIGNAL
C     NCS              NUMBER OF CARDS PER SUBJECT
C     NS               NUMBER OF SUBJECTS
C     NV               NUMBER OF VARIABLES
C     SIGMA            STANDARD DEVIATION (POPULATION FORMULA)
C     NT               N UPON WHICH XBAR AND SIGMA ARE BASED
C     SUMX             SUM OF SCORES
C     SUMX2            SUM OF SQUARED SCORES
C     XBAR             MEAN
C
C     ARRAY AND DIMENSIONS              MEANING
C
C       INPUT(80*NCS)            TEMPORARY STORAGE FOR EACH
C                                    SUBJECT'S DATA CARDS
C       KEY(11)                  CHARACTERS BLANK AND 0 THROUGH 9
C       KFREQ(NV,12)             FREQUENCIES OF OCCURRENCE OF
C                                    DATA CODES
C       KOLS(80*NCS)             VARIABLE LOCATION CARD(S)
C                                    SPECIFICATIONS
C       KP(20)                   PARAMETERS RETURNED BY START
C       TEMP(6)                  TEMPORARY STORAGE
C
C     DISK FILE 1 (NV RECORDS BY 20 WORDS PER RECORD) IS USED FOR
C       STORAGE OF VARIABLE NAMES.  IF NV DIVIDED BY 4 YIELDS A
C       NON-ZERO REMAINDER, INCREASE NUMBER OF RECORDS IN FILE BY 2.
C
C     THIS ROUTINE TESTED UNDER VERSION 2, MODIFICATION LEVEL 7, OF THE
C       IBM 1130 DISK OPERATING SYSTEM ON AN 8-K SYSTEM.
C
```

Figure 8-14. Program D03, System Control Cards and Comment Cards

PROGRAM LOGIC

The source statements in Figure 8-15 round out the rest of the mainline D03. As in the mainline C01, comment cards are used in an effort to provide a clear explanation of the source code.

```
C
C       D03 -- BODY OF PROGRAM
C
        DIMENSION KP(20)
        COMMON KARDS, LINES, NAMES, NCS, NS, NV, INPUT(400),
      1 KFREQ(100,12), KOLS(400)
        DEFINE FILE 1 (100,20,U,N1)
C
C       DEFINE I/O DEVICE NUMBERS.
C
        KARDS = 2
        LINES = 5
C
C       PRINT PROGRAM IDENTIFICATION.
C
        WRITE (LINES,5)
      5 FORMAT (81H1PROGRAM D03, FREQUENCY TABULATION FOR SINGLE-COLUMN VA
      1RIABLES, VERSION OF 1/1/72 / 61H GEORGE PEABODY COLLEGE COMPUTER C
      2ENTER, NASHVILLE, TENNESSEE /)
C
C       INPUT TITLE AND PARAMETER CARDS.
C
        CALL START (KP,KARDS,LINES)
C
C       DEFINE PARAMETERS.
C
        NV = KP(1)
        NS = KP(2)
        NAMES = KP(3)
        NCS = KP(4)
C
C       ECHO PARAMETERS.
C
        WRITE (LINES,10) NV, NS, NAMES, NCS
     10 FORMAT (23HOPARAMETER CARD ENTRIES / 24H COLUMNS  ENTRY  MEANING /
      1 6H0  1-4, I8, 3X, 19HNUMBER OF VARIABLES / 6H  5-8, I8, 3X,
      2 18HNUMBER OF SUBJECTS / 7H  9-12, I7, 3X,
      3 21HVARIABLE NAMES OPTION /7H  13-16, I7, 3X,
      4 27HNUMBER OF CARDS PER SUBJECT)
C
C       TEST VALIDITY OF PARAMETERS.
C
        IF (NV) 15,15,25
     15 WRITE (LINES,20)
     20 FORMAT (26HOZERO VARIABLES SPECIFIED.)
        CALL EXIT
     25 IF (NV - 100) 40,40,30
     30 WRITE (LINES,35)
     35 FORMAT (32HOTOO MANY VARIABLES (MAX = 100).)
        CALL EXIT
     40 IF (NS) 45,45,55
     45 WRITE (LINES,50)
     50 FORMAT (25HOZERO SUBJECTS SPECIFIED.)
        CALL EXIT
     55 IF (NCS) 60,60,70
     60 WRITE (LINES,65)
     65 FORMAT (34HOZERO CARDS PER SUBJECT SPECIFIED.)
        CALL EXIT
     70 IF (NCS - 5) 85,85,75
     75 WRITE (LINES,80)
     80 FORMAT (38HOTOO MANY CARDS PER SUBJECT (MAX = 5).)
        CALL EXIT
C
C       INPUT VARIABLE LOCATION CARD(S), CHECKING FOR ERRORS.
C
     85 NSPEC = 0
        DO 115 I = 1,NCS
        READ (KARDS,90) (INPUT(J), J = 1,80)
     90 FORMAT (80A1)
        NE = 80 * (I - 1)
        DO 115 J = 1,80
        NE = NE + 1
        CALL DCODE (INPUT,J,J,KOLS(NE),IER)
        IF (IER) 95,95,105
     95 IF (KOLS(NE) - 1) 115,100,105
    100 NSPEC = NSPEC + 1
        GO TO 115
    105 WRITE (LINES,110) J, (INPUT(K), K = 1,80)
    110 FORMAT (24HOINVALID ENTRY IN COLUMN, I3, 49H OF PRESUMED VARIABLE
      1LOCATION CARD ON NEXT LINE. / 1X, 80A1)
        CALL EXIT
    115 CONTINUE
        IF (NV - NSPEC) 120,130,120
    120 WRITE (LINES,125)
    125 FORMAT (74HONUMBER OF ENTRIES ON VARIABLE LOCATION CARD(S) DOES NO
      1T AGREE WITH NUMBER / 24H OF VARIABLES SPECIFIED.)
        CALL EXIT
```

Figure 8-15. Body of Program D03

```
C
C      INPLT VARIABLE NAMES, IF CPTEC.
C
   130 IF (NAMES) 145,145,135
   135 CC 140 I = 1,NV,4
       READ (KARCS,90) (INPUT(J), J = 1,80)
   140 WRITE (1'I) (INPUT(J), J = 1,80)
C
C      TRANSFER CCNTRCL TC NEXT SEGMENT.
C
   145 CALL LINK (D03A)
       END
// DUP
*STORECI    WS  LA  C03
```

Figure 8-15. (continued)

The flow of program logic is very similar to that in C01, except for the statements devoted to input, echo, and checking of the variable location card(s). Statements 85 through 125 correspond to the call to subroutine FMAT in C01. Because the structure of this sequence of statements is based on a similar sequence and remarks made in the introduction of variable location cards in Chapter 2, no analysis of statements 85 through 125 is presented here.

The remaining sections of D03 are again very similar to the corresponding sections in C01. In fact, an inspection of both C01 and D03 leads to the conclusion that the transfer of techniques from one library program to another is as significant for the programmer as it is for the user.

As mentioned earlier, no subprograms are LOCALed in the core load for D03. Accordingly, no supervisor control record follows the *STORECI card, and columns 29 and 30 of the same card are left blank.

Program D03A: Data Input and Output of Results

Figure 8-16 presents a highly simplified flow chart for D03A. Because disk input of data is not an option available with program D03, the complexity of D03A is considerably less than that of the three mainlines C01A, C01B, and C01C.

INPUT OF DATA

As seen in Figure 8-17, the D03A source statements devoted to input of data are quite different from the corresponding statements in C01A. The use of variable location cards, differentiation between zero and blank, and acceptance of nonnumeric, nonblank codes requires the use of character manipulation in a way not seen in C01A.

After each data card is read using subroutine READ, the appropriate variable location specifications stored in the array KOLS are scanned to ascertain from which columns (if any) of that data card codes are to be taken for tabulation in KFREQ, the array of frequencies. The array KFREQ is dimensioned with rows corresponding to variables and with columns corresponding to the

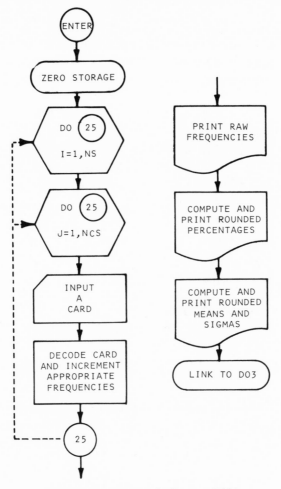

Figure 8-16. Flow Chart for Program D03A

codes blank, 0 through 9, and "other," in that order. Each code read is compared to the elements of the array KEY (defined in a DATA statement) to find a match. If a match is found, L, the corresponding column of KFREQ is simply defined as the number of the matching KEY element. If no match is found, the code is tabulated in column 12, "other."

OUTPUT OF RAW FREQUENCIES

Rather than two separate sequences for output with or without variable names, a compromise is adopted in D03A. A heading including the legend "DESCRIPTION (IF ANY)" is printed, and one of two WRITE statements is

```
// *
// *   D03A -- FIRST PART
// *
// FOR      PROGRAM D03A (LINK FROM D03)
*ONE WORD INTEGERS
*LIST SOURCE PROGRAM
*IOCS(1403PRINTER,DISK)
C         PROGRAM D03A, LINK FROM PROGRAM D03
C
C         THIS SEGMENT INPUTS DATA, COMPUTES STATISTICS, AND PRINTS RESULTS.
C
C         GEORGE PEABODY COLLEGE COMPUTER CENTER, NASHVILLE, TENNESSEE
C         JAMES H. HOGGE, 1/1/72
C
          DIMENSION TEMP(6), KEY(11)
          COMMON KARDS, LINES, NAMES, NCS, NS, NV, INPUT(400),
         1 KFREQ(100,12), KOLS(400)
C
C         SET UP INTEGER REPRESENTATIONS OF BLANK AND NUMERIC CHARACTERS.
C
          DATA KEY /' ','0','1','2','3','4','5','6','7','8','9'/
C
          DEFINE FILE 1 (100,20,L,N1)
C
C         ZERO STORAGE.
C
          DO 5 J = 1,12
          DO 5 I = 1,NV
        5 KFREQ(I,J) = 0
C
C         INPUT DATA AND ACCUMULATE FREQUENCIES.
C
          DO 25 I = 1,NS
          NSPEC = 0
          DO 25 J = 1,NCS
          NSTOP = J * 80
          NSTAR = NSTOP - 79
C
C         INPUT ONE DATA CARD.
C
          CALL READ (INPUT,NSTAR,NSTOP,NER)
C
C         SCAN VARIABLE LOCATION SPECIFICATIONS FOR COLUMNS TO BE READ.
C
          DO 25 K = NSTAR,NSTOP
          IF (KOLS(K)) 25,25,10
       10 NSPEC = NSPEC + 1
C
C         INSPECT CODE AND INCREMENT APPROPRIATE CELL OF KFREQ.
C
          DO 15 L = 1,11
          IF (INPUT(K) - KEY(L)) 15,20,15
       15 CONTINUE
          L = 12
       20 KFREQ(NSPEC,L) = KFREQ(NSPEC,L) + 1
       25 CONTINUE
C
C         PRINT HEADINGS AND TABLE OF FREQUENCIES.
C
          II = 0
          WRITE (LINES,30) (I, I = II,9)
       30 FORMAT (40HOTABLE OF FREQUENCIES (ROWS = VARIABLES) // 5X,
         1 5HBLANK, I3, 9I4, 29H  OTHER  DESCRIPTION (IF ANY) /)
          DO 50 I = 1,NV
C
C         CHECK FOR VARIABLE NAMES, AND BRANCH ACCORDINGLY.
C
          IF (NAMES) 45,45,35
C
C         READ VARIABLE NAME FROM DISK.
C
       35 READ (1'I) (INPUT(J), J = 1,20)
C
C         PRINT WITH VARIABLE NAME.
C
          WRITE (LINES,40) I, (KFREQ(I,J), J = 1,12), (INPUT(K), K = 1,20)
       40 FORMAT (I4, I5, 10I4, I5, 4X, 20A1)
          GO TO 50
C
C         PRINT WITHOUT VARIABLE NAME.
C
       45 WRITE (LINES,40) I, (KFREQ(I,J), J = 1,12)
       50 CONTINUE
```

Figure 8-17. First Part of Program D03A (Link from D03)

used to print each row of the table, with or without variable names. The two WRITE statements may share the same FORMAT statement because the variable name, if present, appears at the end of each line.

OUTPUT OF TABLES OF PERCENTAGES

Figure 8-18 presents the remaining source statements from D03A. In order to stay within a limit of 81 print positions (to allow binding with 8-1/2 by 11 inch sheets and maintain compatibility with a popular line printer featured in small-scale computer systems), percentages are printed in two sections. Subroutine ROUND is used for rounding, and the strategy of two WRITE statements referencing the same FORMAT statement is used to accommodate the variable names option.

OUTPUT OF MEANS AND SIGMAS

Included in this output is an echo of the variable location card entries. The values ΣX and ΣX^2 are computed by multiplying each possible nonzero score (or its square) by its frequency of occurrence. After an unrounded mean is computed and used in the computation of a rounded sigma, the mean is rounded for printing. The variable names option is handled as in the previous two sections of output.

Finally, the CALL LINK statement is used to transfer control back to the first segment, D03.

Evidence of the relative simplicity of D03A as compared to C01A, D01B, and C01C is again seen in the system control cards, where only two small subprograms, READ and ROUND, are LOCALed.

SUMMARY

This chapter represents an attempt to serve the integrative function of presenting two library programs (or systems of segments, as we have seen) incorporating many of the techniques and standards discussed throughout this volume. It is hoped that the reader has now reached the point where he can begin the development of user-oriented library programs adapted to his own particular biases and operating environment.

```
C
C       CC3A -- SECOND PART
C
C
C       PRINT HEADINGS AND TABLES CF PERCENTACES.
C
        WRITE (LINES,55) (I, I = II,4)
     55 FCRMAT (22HOTABLES OF PERCENTAGES // 16H VARIABLE   BLANK, I6,
       1 4I8, 5X, 20HCESCRIPTICN (IF ANY) /)
        T = NS
        CC 80 I = 1,NV
C
C       CCMPUTE RCUNDEC PERCENTAGES.
C
        CC 60 J = 1,6
     60 CALL RCUNC (FLCAT(KFREC(I,J))/T*100.0,2,TEMP(J))
C
C       CHECK FCR VARIABLE NAMES, ANC BRANCH ACCORDINGLY.
C
        IF (NAMES) 75,75,65
C
C       READ VARIABLE NAME FROM CISK.
C
     65 READ (1'I) (INPUT(J), J = 1,20)
C
C       PRINT WITH VARIABLE NAME.
C
        WRITE (LINES,70) I, TEMP, (INPUT(J), J = 1,20)
     70 FCRMAT (I6, 2X, 6F8.2, 3X, 20A1)
        GO TO 80
C
C       PRINT WITHCUT VARIABLE NAME.
C
     75 WRITE (LINES,7C) I, TEMP
     80 CONTINUE
C
C       PRINT TABLE FCR REMAINING CATEGORIES.
C
        WRITE (LINES,85) (I, I = 5,9)
     85 FCRMAT (9HOVARIABLE, I5, 4I8, 5X, 28HOTHER   DESCRIPTION (IF ANY)
       1 /)
        CC 105 I = 1,NV
C
C       CCMPUTE RCUNDEC PERCENTAGES.
C
        CC 90 J = 1,6
     90 CALL RCUNC (FLCAT(KFREC(I,J+6))/T*100.0,2,TEMP(J))
C
C       CHECK FCR VARIABLE NAMES, ANC BRANCH ACCORDINGLY.
C
        IF (NAMES) 100,100,95
C
C       READ VARIABLE NAME FRCM CISK.
C
     95 READ (1'I) (INPUT(J), J = 1,20)
C
C       PRINT WITH VARIABLE NAME.
C
        WRITE (LINES,70) I, TEMP, (INPUT(J), J = 1,20)
        GC TO 105
C
C       PRINT WITHCUT VARIABLE NAME.
C
    100 WRITE (LINES,70) I, TEMP
    105 CONTINUE
C
C       COMPUTE MEANS AND SIGMAS, ANC ECHO VARIABLE LOCATION
C           SPECIFICATIONS.
C
C       PRINT HEADINGS.
C
        WRITE (LINES,110)
    110 FORMAT (63HOMEAN, SIGMA, AND LOCATION OF EACH VARIABLE IN SUBJECT
       1CARD SET / 62H (BLANK, 0, AND OTHER EXCLUDED FROM COMPUTATION OF S
       2TATISTICS) / 70HOVARIABLE   CARD COLUMN      MEAN      SIGMA
       3CESCRIPTICN (IF ANY) /)
C
        NE = 0
        CC 145 I = 1,NV
C
C       SEARCH FCR NEXT VARIABLE LCCATION SPECIFICATION.
C
    115 NE = NE + 1
        IF (KCLS(NE)) 115,115,120
```

Figure 8-18. Second Part of Program D03A

```
C
C      COMPUTE CARD (II) AND COLUMN (JJ).
C
  120 II = NE / 80 + 1
      JJ = NE - (II - 1) * 80
C
C      INITIALIZE ACCUMULATORS.
C
      NT = 0
      SUMX = 0.0
      SUMX2 = 0.0
C
C      COMPUTE MEAN (XBAR) AND SIGMA.
C
      DO 125 J = 1,9
      F = KFREQ(I,J+2)
      XJ = J
      NT = NT + KFREQ(I,J+2)
      SUMX = SUMX + F * XJ
  125 SUMX2 = SUMX2 + F * XJ * XJ
      TN = NT
C
C      COMPUTE MEAN WITHOUT ROUNDING, COMPUTE ROUNDED SIGMA, THEN
C          ROUND MEAN.
C
      XBAR = SUMX / TN
      CALL ROUND (SQRT(SUMX2/TN-XBAR*XBAR),3,SIGMA)
      CALL ROUND (XBAR,3,XBAR)
C
C      CHECK FOR VARIABLE NAMES AND BRANCH ACCORDINGLY.
C
      IF (NAMES) 140,140,130
C
C      READ VARIABLE NAME FROM DISK.
C
  130 READ (1'I) (INPUT(J), J = 1,20)
C
C      PRINT WITH VARIABLE NAME.
C
      WRITE (LINES,135) I, II, JJ, XBAR, SIGMA, (INPUT(J), J = 1,20)
  135 FORMAT (I6, I8, I7, F13.3, F11.3, 5X, 20A1)
      GO TO 145
C
C      PRINT WITHOUT VARIABLE NAME.
C
  140 WRITE (LINES,135) I, II, JJ, XBAR, SIGMA
  145 CONTINUE
C
C      TRANSFER CONTROL TO FIRST SEGMENT TO LOOK FOR AN ADDITIONAL
C          DATA SET.
C
      CALL LINK (DO3)
      END
// DUP
*STORECI     WS  LA  CO3A      1
*LOCAL,READ,ROUND
```

Figure 8-18. (continued)

Appendix A:
Listings of
Subprograms

This Appendix contains source listings for the subprograms described in the body of this book. All routines have been written in IBM 1130 FORTRAN.

The conversion of certain subprograms for use on a given computer system may require modification of the DATA statements, since the syntax associated with alphanumeric constants varies from machine to machine. On some computers FORTRAN lacks DATA statements altogether; nevertheless, it is still possible to define alphanumeric constants through the use of the integer equivalents for the characters involved. Although these equivalents differ from computer to computer, you can discover them for your computer with the program in Figure A-1. The single data card read by this program will contain, beginning in column 1,

ABCDEFGHIJKLMNOPQRSTUVWXYZ1234567890=b/,()-$+.*'

(where b = blank). The program prints the integer equivalents (to the base 10) of the above characters, stored one character per integer variable (or, more accurately, integer array element). If your computer's FORTRAN does not recognize DATA statements, you can utilize arithmetic statements to define alphanumeric constants. For example,

DATA KBLNK /' '/

may be replaced by

KBLNK = 16448

if 16448 is the integer equivalent of a blank.

```
C
C       ROUTINE TO PRINT INTEGER EQUIVALENTS OF ALPHAMERIC CHARACTERS
C
        DIMENSION KCHAR(48)
C
C       KARDS = DEVICE NUMBER FOR CARD READER.
C       LINES = DEVICE NUMBER FOR LINE PRINTER.
C
        KARDS = 2
        LINES = 5
C
C       INPUT FORTRAN CHARACTER SET.
C
        READ (KARDS,5) (KCHAR(I), I = 1,48)
      5 FORMAT (48A1)
C
C       PRINT HEADINGS.
C
        WRITE (LINES,10)
     10 FORMAT (5X, 9HCHARACTER, 3X, 4HCODE)
        DO 15 I = 1,48
C
C       PRINT A CHARACTER AND ITS INTEGER EQUIVALENT.
C
     15 WRITE (LINES,20) KCHAR(I), KCHAR(I)
     20 FORMAT (9X, A1, I11)
        CALL EXIT
        END
```

Figure A-1. Routine to Print Integer Equivalents of Alphanumeric Characters

```
        SUBROUTINE DCODE (KK,NSTAR,NSTOP,INTEG,IER)
C
C       ROUTINE TO EXTRACT INTEGER FROM NUMERIC CHARACTERS
C
C       KK = ONE-DIMENSIONAL INTEGER ARRAY WHICH CONTAINS, IN A1 FORMAT,
C           THE NUMERIC CHARACTERS TO BE DECODED.
C       NSTAR = LEFTMOST POSITION (LOWEST-NUMBERED ELEMENT) IN KK TO BE
C           DECODED.
C       NSTOP = RIGHTMOST POSITION (HIGHEST-NUMBERED ELEMENT) IN KK TO BE
C           DECODED.
C       INTEG = INTEGER VARIABLE WHICH WILL CONTAIN THE RESULT DECODED.
C       IER = INTEGER VARIABLE WHICH WILL EQUAL ZERO UPON RETURN IF ONLY
C           BLANKS OR NUMERIC CHARACTERS ARE ENCOUNTERED IN THE FIELD.
C           IF A NON-NUMERIC, NON-BLANK CHARACTER IS ENCOUNTERED, IER
C           WILL CONTAIN THE NUMBER OF THE ELEMENT OF KK IN WHICH THE
C           ERROR OCCURRED, AND INTEG WILL BE ZERO.  THEREFORE, IER
C           SHOULD BE TESTED AFTER EACH CALL TO DCODE.
C
C       JAMES H. HOGGE, GEORGE PEABODY COLLEGE, NASHVILLE, TENNESSEE
C       VERSION OF 1/1/72
C
        DIMENSION KK(1), KEY(10)
        DATA KBLNK, KEY /' ','0','1','2','3','4','5','6','7','8','9'/
        INTEG = 0
        IER = 0
        KOUNT = 0
C
C       BEGIN SCAN OF FIELD.
C
        DO 20 I = NSTAR,NSTOP
        KOUNT = KOUNT + 1
C
C       CHECK FOR BLANK.
C
        IF (KK(I) - KBLNK) 10,5,10
      5 J = 1
        GO TO 20
C
C       SEARCH FOR MATCHING NUMERAL.
C
     10 DO 15 J = 1,10
        IF (KK(I) - KEY(J)) 15,20,15
     15 CONTINUE
C
C       NO MATCH.  SET ERROR INDICATOR AND RETURN.
C
        IER = I
        RETURN
C
C       INCREMENT INTEG BY NUMBER * 10 TO APPROPRIATE POWER.
C
     20 INTEG = INTEG + (J - 1) * 10 ** (NSTOP - NSTAR + 1 - KOUNT)
        RETURN
        END
```

```
      SUBROUTINE DCPT (KFILE,N,LINES)
C
C     ROUTINE TO PRINT A DISK-RESIDENT COMPRESSED SYMMETRIC MATRIX
C
C     KFILE = NUMBER OF DISK FILE IN WHICH THE UPPER TRIANGLE OF THE
C             MATRIX, INCLUDING THE MAIN DIAGONAL, IS STORED AS A
C             SERIES OF TWO-WORD (REAL) RECORDS.
C     N = NUMBER OF ROWS AND COLUMNS IN MATRIX.
C     LINES = OUTPUT DEVICE NUMBER.
C
C     THIS ROUTINE REFERENCES SUBROUTINE ROUND.
C
C     JAMES H. HOGGE, GEORGE PEABODY COLLEGE, NASHVILLE, TENNESSEE
C     VERSION OF 1/1/72
C
      DIMENSION Y(8)
      NTOT = (N * N - N) / 2 + N
C
C     PARTITION MATRIX INTO 8-COLUMN BLOCKS.
C
      DO 45 I = 1,N,8
      IF (I + 7 - N) 5,10,10
    5 J = I + 7
      GO TO 15
   10 J = N
   15 WRITE (LINES,20) (K, K = I,J)
   20 FORMAT (/ 2X, 8I9)
      DO 45 L = 1,N
      NL = 0
      DO 40 K = I,J
      NL = NL + 1
      IF (L - K) 25,25,30
   25 LL = L
      KK = K
      GO TO 35
   30 LL = K
      KK = L
   35 NE = NTOT - N + KK - (N - LL) * (N - LL + 1) / 2
C
C     READ ELEMENT FROM DISK.
C
      READ (KFILE'NE) Y(NL)
C
C     ROUND FOR PRINTING.
C
   40 CALL ROUND (Y(NL),3,Y(NL))
   45 WRITE (LINES,50) L, (Y(K), K = 1,NL)
   50 FORMAT (I5, 8F9.3)
      RETURN
      END

      SUBROUTINE DCPTN (KFILE,N,LINES,NFILE)
C
C     ROUTINE TO PRINT A DISK-RESIDENT COMPRESSED SYMMETRIC MATRIX WITH
C         VARIABLE NAMES
C
C     KFILE = NUMBER OF DISK FILE IN WHICH THE UPPER TRIANGLE OF THE
C             MATRIX, INCLUDING THE MAIN DIAGONAL, IS STORED AS A
C             SERIES OF TWO-WORD (REAL) RECORDS.
C     N = NUMBER OF ROWS AND COLUMNS IN MATRIX.
C     LINES = OUTPUT DEVICE NUMBER.
C     NFILE = NUMBER OF DISK FILE IN WHICH VARIABLE NAMES ARE STORED,
C             20 CHARACTERS PER NAME, ONE NAME PER RECORD, ONE
C             CHARACTER PER WORD.
C
C     THIS ROUTINE REFERENCES SUBROUTINE ROUND.
C
C     JAMES H. HOGGE, GEORGE PEABODY COLLEGE, NASHVILLE, TENNESSEE
C     VERSION OF 1/1/72
C
      DIMENSION Y(6), NAME(20)
      NTOT = (N * N - N) / 2 + N
C
C     PARTITION MATRIX INTO 6-COLUMN BLOCKS.
C
      DO 45 I = 1,N,6
      IF (I + 5 - N) 5,10,10
    5 J = I + 5
      GO TO 15
   10 J = N
   15 WRITE (LINES,20) (K, K = I,J)
   20 FORMAT (/ 23X, 6I9)
      DO 45 L = 1,N
      NL = 0
C
```

```
      C     READ VARIABLE NAME FROM CISK.
      C
            READ (NFILE'L) NAME
            CC 40 K = I,J
            NL = NL + 1
            IF (L - K) 25,25,30
         25 LL = L
            KK = K
            GC TO 35
         30 LL = K
            KK = L
         35 NE = NTOT - N + KK - (N - LL) * (N - LL + 1) / 2
      C
      C     READ ELEMENT FRCM DISK.
      C
            READ (KFILE'NE) Y(NL)
      C
      C     RCUND FOR PRINTING.
      C
         40 CALL ROUND (Y(NL),3,Y(NL))
         45 WRITE (LINES,50) L, NAME, (Y(K), K = 1,NL)
         50 FORMAT (I5, 1X, 20A1, 6F5.3)
            RETURN
            END

            SUBROUTINE DDPCH (KFILE,N,KPNCH,MATID)
      C
      C     ROUTINE TC PUNCH THE UPPER TRIANGLE OF A DISK-RESIDENT COMPRESSED
      C        SYMMETRIC MATRIX
      C
      C     KFILE = NUMBER OF DISK FILE IN WHICH THE UPPER TRIANGLE OF THE
      C             MATRIX, INCLUDING THE MAIN DIAGONAL, IS STORED AS A
      C             SERIES OF TWO-WORD (REAL) RECORDS.
      C     N = NUMBER OF ROWS AND COLUMNS IN MATRIX.
      C     KPNCH = DEVICE NUMBER FOR CARD PUNCH.
      C     MATID = FIVE-ELEMENT INTEGER ARRAY CONTAINING IDENTIFICATION
      C             FIELD FOR PUNCHED OUTPUT, ONE CHARACTER PER ELEMENT.
      C
      C     THIS VERSION WRITTEN FCR LSE WITH IBM 1442 CARD READ/PUNCH UNIT.
      C     SEE COMMENTS BELOW.
      C
      C     THIS ROUTINE REFERENCES SUBROUTINE RCUND.
      C
      C     JAMES H. HOGGE, GEORGE PEABODY COLLEGE, NASHVILLE, TENNESSEE
      C     VERSION CF 1/1/72
      C
            DIMENSION MATID(5), Y(7)
      C
      C     THE FOLLCWING SET OF STATEMENTS MUST BE REMOVED IF THIS ROLTINE
      C        IS TO BE USED ON A SYSTEM WITH SEPARATE DEVICES FOR READING
      C        AND PLNCHING CARDS.
      C
            DIMENSION KARD(40)
      C
      C     SET UP BLANKS FCR LATER LSE.
      C
            DATA KBLNK /' '/
      C
      C     END OF FIRST SET OF STATEMENTS TO BE REMOVED.
      C
      C     CCMPUTE TOTAL NUMBER OF ELEMENTS IN X.
      C
            NTCT = (N * N - N) / 2 + N
      C
      C     INITIALIZE SEQUENCE NUMBER.
      C
            NSEQ = 1
      C
            DO 40 I = 1,NTCT,7
      C
      C     THE FOLLCWING SET OF STATEMENTS MUST BE REMOVED IF THIS ROLTINE
      C        IS TO BE USED ON A SYSTEM WITH SEPARATE DEVICES FOR READING
      C        AND PUNCHING CARDS.
      C
      C     INPUT A CARD.
      C
            READ (KPNCH,5) KARD
          5 FORMAT (40A2)
      C
      C     RETURN WITHOUT PUNCHING IF CARD IS NOT COMPLETELY BLANK.
      C
            CC 10 J = 1,40
            IF (KARD(J) - KBLNK) 45,10,45
         10 CONTINUE
      C
      C     END OF SECOND SET OF STATEMENTS TO BE REMOVED.
```

```
C
C          SEE IF END OF ARRAY HAS BEEN REACHED, AND SET LAST ELEMENT TO
C              BE PUNCHED ACCORDINGLY.
       IF (I + 6 - NTOT) 15,15,20
    15 J = I + 6
       GO TO 25
    20 J = NTOT
C
C          READ BLOCK OF ELEMENTS FROM DISK.
C
    25 NE = J - I + 1
       READ (KFILE'I) (Y(K), K = 1,NE)
C
C          ROUND BEFORE PUNCHING.
C
       DO 30 K = 1,NE
    30 CALL ROUND (Y(K),4,Y(K))
C
C          PUNCH BLOCK OF ELEMENTS.
C
       WRITE (KPNCH,35) MATID, NSEQ, (Y(K), K = 1,NE)
    35 FORMAT (5A1, I5, 7F10.4)
C
C          INCREMENT SEQUENCE NUMBER.
C
    40 NSEQ = NSEQ + 1
    45 RETURN
       END

       SUBROUTINE OPNCH (X,N,KPNCH,MATID)
C
C          ROUTINE TO PUNCH THE UPPER TRIANGLE OF A COMPRESSED SYMMETRIC
C              MATRIX
C
C          X = ONE-DIMENSIONAL REAL ARRAY CONTAINING THE SYMMETRIC MATRIX
C              IN COMPRESSED FORM (UPPER TRIANGLE, INCLUDING MAIN DIAGONAL).
C          N = NUMBER OF ROWS AND COLUMNS IN MATRIX REPRESENTED IN X.
C          KPNCH = DEVICE NUMBER FOR CARD PUNCH.
C          MATID = FIVE-ELEMENT INTEGER ARRAY CONTAINING IDENTIFICATION
C              FIELD FOR PUNCHED OUTPUT, ONE CHARACTER PER ELEMENT.
C
C          THIS VERSION WRITTEN FOR USE WITH IBM 1442 CARD READ/PUNCH UNIT.
C          SEE COMMENTS BELOW.
C
C          JAMES H. HOGGE, GEORGE PEABODY COLLEGE, NASHVILLE, TENNESSEE
C          VERSION OF 1/1/72
C
       DIMENSION X(1), MATID(5)
C
C          THE FOLLOWING SET OF STATEMENTS MUST BE REMOVED IF THIS ROUTINE
C              IS TO BE USED ON A SYSTEM WITH SEPARATE DEVICES FOR READING
C              AND PUNCHING CARDS.
C
       DIMENSION KARD(40)
C
C          SET UP BLANKS FOR LATER USE.
C
       DATA KBLNK /'  '/
C
C          END OF FIRST SET OF STATEMENTS TO BE REMOVED.
C
C          COMPUTE TOTAL NUMBER OF ELEMENTS IN X.
C
       NTOT = (N * N - N) / 2 + N
C
C          INITIALIZE SEQUENCE NUMBER.
C
       NSEQ = 1
C
       DO 35 I = 1,NTOT,7
C
C          THE FOLLOWING SET OF STATEMENTS MUST BE REMOVED IF THIS ROUTINE
C              IS TO BE USED ON A SYSTEM WITH SEPARATE DEVICES FOR READING
C              AND PUNCHING CARDS.
C
C          INPUT A CARD.
C
       READ (KPNCH,5) KARD
     5 FORMAT (40A2)
C
C          RETURN WITHOUT PUNCHING IF CARD IS NOT COMPLETELY BLANK.
C
       DO 10 J = 1,40
       IF (KARD(J) - KBLNK) 40,10,40
    10 CONTINUE
C
```

```
C     END OF SECOND SET OF STATEMENTS TO BE REMOVED.
C
C     SEE IF END OF ARRAY HAS BEEN REACHED, AND SET LAST ELEMENT TO
C        BE PUNCHED ACCORDINGLY.
      IF (I + 6 - NTCT) 15,15,20
   15 J = I + 6
      GO TO 25
   20 J = NTOT
C
C     PUNCH BLOCK OF ELEMENTS.
C
   25 WRITE (KPNCH,30) MATID, NSEQ, (X(K), K = I,J)
   30 FORMAT (5A1, I5, 7F10.4)
C
C     INCREMENT SEQUENCE NUMBER.
C
   35 NSEQ = NSEQ + 1
   40 RETURN
      END

      SUBROUTINE DPRT (X,N,LINES)
C
C     ROUTINE TO PRINT A COMPRESSED SYMMETRICAL MATRIX
C
C     X = ONE-DIMENSIONAL REAL ARRAY IN WHICH THE UPPER TRIANGLE,
C        INCLUDING THE MAIN DIAGONAL, OF A SYMMETRICAL MATRIX IS
C        STORED IN COMPRESSED FORM.
C     N = NUMBER OF ROWS AND COLUMNS IN MATRIX REPRESENTED IN ARRAY X.
C     LINES = OUTPUT DEVICE NUMBER.
C
C     JAMES H. HOGGE, GEORGE PEABODY COLLEGE, NASHVILLE, TENNESSEE
C     VERSION OF 1/1/72
C
      DIMENSION X(1), Y(8)
      NTOT = (N * N - N) / 2 + N
C
C     PARTITION MATRIX INTO 8-COLUMN BLOCKS.
C
      DO 45 I = 1,N,8
      IF (I + 7 - N) 5,10,10
    5 J = I + 7
      GO TO 15
   10 J = N
   15 WRITE (LINES,20) (K, K = I,J)
   20 FORMAT (/ 2X, 8I9)
      DO 45 L = 1,N
      NL = 0
      DO 40 K = I,J
      NL = NL + 1
      IF (L - K) 25,25,30
   25 LL = L
      KK = K
      GO TO 35
   30 LL = K
      KK = L
   35 NE = NTOT - N + KK - (N - LL) * (N - LL + 1) / 2
   40 Y(NL) = X(NE)
   45 WRITE (LINES,50) L, (Y(K), K = 1,NL)
   50 FORMAT (I5, 8F9.3)
      RETURN
      END

      SUBROUTINE DPRTN (X,N,LINES,NFILE)
C
C     ROUTINE TO PRINT A COMPRESSED SYMMETRIC MATRIX WITH VARIABLE
C        NAMES
C
C     X = ONE-DIMENSIONAL REAL ARRAY IN WHICH THE UPPER TRIANGLE,
C        INCLUDING THE MAIN DIAGONAL, OF A SYMMETRICAL MATRIX IS
C        STORED IN COMPRESSED FORM.
C     N = NUMBER OF ROWS AND COLUMNS IN MATRIX REPRESENTED IN ARRAY X.
C     LINES = OUTPUT DEVICE NUMBER.
C     NFILE = NUMBER OF DISK FILE IN WHICH VARIABLE NAMES ARE STORED,
C        20 CHARACTERS PER NAME, ONE NAME PER RECORD, ONE
C        CHARACTER PER WORD.
C
C     JAMES H. HOGGE, GEORGE PEABODY COLLEGE, NASHVILLE, TENNESSEE
C     VERSION OF 1/1/72
C
      DIMENSION X(1), Y(6), NAME(20)
      NTOT = (N * N - N) / 2 + N
C
C     PARTITION MATRIX INTO 6-COLUMN BLOCKS.
C
      DO 45 I = 1,N,6
      IF (I + 5 - N) 5,10,10
```

```
      5 J = I + 5
        GO TO 15
     10 J = N
     15 WRITE (LINES,20) (K, K = I,J)
     20 FORMAT (/ 23X, 6I9)
        DO 45 L = 1,N
        NL = 0
C
C       READ VARIABLE NAME FROM DISK.
C
        READ (NFILE'L) NAME
        DO 40 K = I,J
        NL = NL + 1
        IF (L - K) 25,25,30
     25 LL = L
        KK = K
        GO TO 35
     30 LL = K
        KK = L
     35 NE = NTOT - N + KK - (N - LL) * (N - LL + 1) / 2
     40 Y(NL) = X(NE)
     45 WRITE (LINES,50) L, NAME, (Y(K), K = 1,NL)
     50 FORMAT (I5, 1X, 20A1, 6F9.3)
        RETURN
        END

        SUBROUTINE ECODE (KK,NSTAR,NSTOP,INTEG)
C
C       ROUTINE TO CONVERT INTEGER TO CHARACTERS IN AN ARRAY
C
C       KK = ONE-DIMENSIONAL INTEGER ARRAY INTO WHICH INTEG IS TO BE
C            PLACED AS CHARACTERS, ONE DIGIT PER ELEMENT.
C       NSTAR = FIRST ELEMENT OF KK TO BE FILLED.
C       NSTOP = LAST ELEMENT OF KK TO BE FILLED.
C       INTEG = NUMBER TO BE CONVERTED TO CHARACTERS AND PLACED IN KK.
C
C       THE CHARACTERS REPRESENTING INTEG ARE RIGHT-JUSTIFIED IN KK,
C            LEADING ZEROS APPEAR AS BLANKS, AND A MINUS SIGN WILL BE
C            PLACED IN KK, IF NEEDED.  IF INTEG IS TOO BIG TO GO IN
C            THE ELEMENTS ALLOTTED, ASTERISKS WILL BE PLACED IN THE
C            ELEMENTS SPECIFIED.
C
C       JAMES H. HOGGE, GEORGE PEABODY COLLEGE, NASHVILLE, TENNESSEE
C       VERSION OF 1/1/72
C
        DIMENSION KK(1), KDGTS(5), KEY(10)
        DATA KEY /'0','1','2','3','4','5','6','7','8','9'/
        DATA MINUS, KASTR, KBLNK /'-','*',' '/
C
C       REMOVE POSSIBLE NEGATIVE SIGN.
C
        NUM = IABS(INTEG)
C
C       STORE DIGITS OF NUM IN KDGTS.
C
        DO 5 I = 1,5
        KP = 10 ** (5 - I)
        KDGTS(I) = NUM / KP
      5 NUM = NUM - KDGTS(I) * KP
C
C       FIND LAST LEADING ZERO.
C
        DO 10 I = 1,5
        IF (KDGTS(I)) 15,10,15
     10 CONTINUE
        I = 5
C
C       IF THERE IS ROOM, PLACE CONVERTED DIGITS IN KK.
C
     15 ND = 6 - I
        IF (NSTOP - NSTAR + 1 - ND) 60,20,20
     20 JSTAR = NSTOP - ND + 1
        NE = 5
        DO 25 I = JSTAR,NSTOP
        II = KDGTS(NE) + 1
        JJ = NSTOP - I + JSTAR
        KK(JJ) = KEY(II)
     25 NE = NE - 1
C
C       IF INTEG IS NEGATIVE AND THERE IS ROOM, INSERT MINUS SIGN.
C
        IF (INTEG) 30,40,40
     30 ND = ND + 1
        IF (NSTOP - NSTAR + 1 - ND) 60,35,35
     35 NE = NSTOP - ND + 1
        KK(NE) = MINUS
```

```
C
C        BLANK OUT ANY UNUSED ELEMENTS IN KK.
C
      40 NE = NSTOP - ND
         IF (NSTAR - NE) 45,45,55
      45 DO 50 I = NSTAR,NE
      50 KK(I) = KBLNK
      55 RETURN
C
C        NUMBER TOO BIG -- FILL FIELD WITH ASTERISKS.
C
      60 DO 65 I = NSTAR,NSTOP
      65 KK(I) = KASTR
         RETURN
         END

         SUBROUTINE FMAT (KF,KARDS,LINES,LIMIT)
C
C        ROUTINE TO INPUT AND DECODE FORMAT CARD(S)
C
C        KF = ONE-DIMENSIONAL INTEGER ARRAY IN WHICH DECODED FORMAT
C             SPECIFICATIONS ARE TO BE STORED.
C        KARDS = DEVICE NUMBER FOR CARD READER.
C        LINES = DEVICE NUMBER FOR LINE PRINTER.
C        LIMIT = NUMBER OF F FIELDS AND SLASHES TO BE PERMITTED.
C
C        KF MUST BE DIMENSIONED AT LEAST 3 * LIMIT IN CALLING PROGRAM.
C
C        JAMES H. HOGGE, GEORGE PEABODY COLLEGE, NASHVILLE, TENNESSEE
C        VERSION OF 1/1/72
C
         DIMENSION KARD(160), KF(1), KEY(10)
         DATA JLP, JRP, JF, JX, KCMMA, JDEC, KBLNK, KASTR, KEY, KSLSH
        1 /'(',')','F','X','.',',',' ','*','0','1','2','3','4','5','6','7',
        2 '8','9','/'/
         NF = 1
         NT = 3 * LIMIT
         DO 5 I = 1,NT
       5 KF(I) = 0
C
C        INPUT FIRST FORMAT CARD.
C
         READ (KARDS,10) (KARD(I), I = 1,80)
      10 FORMAT (80A1)
         NT = 80
C
C        INPUT SECOND FORMAT CARD IF * IN COLUMN 80.
C
         IF (KARD(80) - KASTR) 20,15,20
      15 READ (KARDS,10) (KARD(I), I = 81,160)
         KARD(80) = KBLNK
         NT = 160
C
C        PRINT ECHO CHECK.
C
      20 WRITE (LINES,25) (KARD(I), I = 1,NT)
      25 FORMAT (17HOFORMAT CARD(S) =/ 2(/1X, 80A1))
C
C        CHECK FOR PARENTHESIS BALANCE.
C
         NL = 0
         NR = 0
         DO 45 I = 1,NT
         IF (KARD(I) - JLP) 35,30,35
      30 NL = NL + 1
         GO TO 45
      35 IF (KARD(I) - JRP) 45,40,45
      40 NR = NR + 1
      45 CONTINUE
         IF (NL - NR) 50,60,50
      50 WRITE (LINES,55)
      55 FORMAT (25HOPARENTHESIS ERROR ABOVE.)
         CALL EXIT
C
C        SEARCH FOR INITIAL LEFT PARENTHESIS.
C
      60 DO 70 I = 1,NT
         IF (KARD(I) - JLP) 70,65,70
      65 JSTAR = I + 1
         GO TO 75
      70 CONTINUE
         GO TO 50
C
C        ESTABLISH INITIAL RIGHT-MOST SCANNING LIMIT.
C
```

```
      75 DO 85 I = JSTAR,NT
         IF (KARD(I) - JLP) 85,80,85
      80 NSTOP = I - 1
         GO TO 90
      85 CONTINUE
         NSTOP = NT - 1
      90 NR = 1
C
C        ENTER MAIN SCANNING LOOP.
C
      95 DO 335 N = 1,NR
         NSTAR = JSTAR
         NUMF = -1
         KWF = 0
         KXF = 0
         KRF = 0
C
C        SEARCH FOR NON-BLANK, NON-NUMERIC CHARACTER.
C
     100 DO 115 I = NSTAR,NSTOP
         IF (KARD(I) - KBLNK) 105,115,105
     105 DO 110 J = 1,10
         IF (KARD(I) - KEY(J)) 110,115,110
     110 CONTINUE
         NN = I - 1
         GO TO 120
     115 CONTINUE
         NN = NSTOP
C
C        EXTRACT NUMBER, IF POSSIBLE.
C
     120 IF (NSTAR - NN) 125,125,160
     125 NUMF = 0
         NDGTS = 0
         DO 150 I = NSTAR,NN
         NE = NN + NSTAR - I
         IF (KARD(NE) - KBLNK) 130,150,130
     130 DO 145 J = 1,10
         IF (KARD(NE) - KEY(J)) 145,135,145
     135 NDGTS = NDGTS + 1
         IF (NDGTS - 4) 140,140,360
     140 NUMF = NUMF + (J - 1) * 10 ** (NDGTS - 1)
         GO TO 150
     145 CONTINUE
         NN = I
         GO TO 360
     150 CONTINUE
         IF (NDGTS) 155,155,160
     155 NUMF = -1
C
C        EXAMINE CHARACTER AND PROCESS ACCORDINGLY.
C
     160 NN = NN + 1
         IF (NN - NSTOP) 165,165,305
     165 IF (KARD(NN) - JX) 190,170,190
     170 IF (KWF) 175,175,360
     175 IF (NUMF) 180,50,185
     180 NUMF = 1
     185 KF(3*NF-2) = KF(3*NF-2) + NUMF
         NUMF = -1
         GO TO 300
     190 IF (KARD(NN) - KOMMA) 225,195,225
     195 IF (KWF) 200,200,205
     200 IF (KRF) 220,220,360
     205 IF (NUMF) 360,210,210
     210 DO 215 I = 1,KRF
         KF(3*NF-1) = KWF
         KF(3*NF) = NUMF
         NF = NF + 1
         IF (NF - LIMIT) 215,380,380
     215 CONTINUE
         KRF = 0
         KWF = 0
         NUMF = -1
         IF (KARD(NN) - KSLSH) 300,270,300
     220 IF (NUMF) 300,360,360
     225 IF (KARD(NN) - JF) 250,230,250
     230 IF (KRF + KWF) 235,235,360
     235 IF (NUMF) 240,360,245
     240 NUMF = 1
     245 KRF = NUMF
         NUMF = -1
         GO TO 300
     250 IF (KARD(NN) - KSLSH) 275,255,275
     255 IF (KWF) 260,260,205
```

```
      260 IF (KRF) 265,265,360
      265 IF (NUMF) 270,360,360
      270 KF(3*NF-1) = -1
          NF = NF + 1
          IF (NF - LIMIT) 300,380,380
      275 IF (KARD(NN) - JDEC) 295,280,295
      280 IF (KRF) 360,360,285
      285 IF (NUMF) 360,360,290
      290 KWF = NUMF
          NUMF = -1
          GO TO 300
      295 IF (KARD(NN) - JRP) 360,305,360
      300 NSTAR = NN + 1
          IF (NSTAR - NSTOP) 100,100,305
C
C         END OF SCANNING LOOP.
C
      305 IF (NUMF) 335,310,310
      310 IF (KWF) 315,315,210
      315 IF (NR - 1) 320,320,360
      320 IF (KARD(NSTOP+1) - JLP) 360,325,360
      325 IF (NUMF) 360,360,330
      330 NR = NUMF
          GO TO 345
      335 CONTINUE
C
C         ESTABLISH NEW SCANNING LIMITS, IF REQUIRED.
C
          IF (NT - NSTOP - 2) 390,340,340
      340 JSTAR = NSTOP + 2
          GO TO 75
      345 JSTAR = NSTOP + 2
          DO 355 I = JSTAR,NT
          IF (KARD(I) - JRP) 355,350,355
      350 NSTOP = I - 1
          GO TO 95
      355 CONTINUE
C
C         ERROR ROUTINE.
C
      360 KK = 1
          IF (NN - 80) 370,370,365
      365 NN = NN - 80
          KK = 2
      370 WRITE (LINES,375) NN, KK
      375 FORMAT (18HOERROR NEAR COLUMN, I3, 15H OF FORMAT CARD, I2)
          CALL EXIT
C
C         OVERFLOW ERROR.
C
      380 WRITE (LINES,385)
      385 FORMAT (33HOFORMAT EXCEEDS PROGRAM CAPACITY.)
          CALL EXIT
C
C         TEST FOR ABSENCE OF F SPECIFICATIONS.
C
      390 KK = 2
      395 IF (KF(KK-1)) 400,405,405
      400 KK = KK + 3
          GO TO 395
      405 IF (KF(KK)) 410,410,420
      410 WRITE (LINES,415)
      415 FORMAT (26HONO F SPECIFICATION ABOVE.)
          CALL EXIT
C
C         TEST FOR SUM OF FIELDS GREATER THAN 80 COLUMNS.
C
      420 KK = 0
          DO 435 I = 1,LIMIT
          IF (KF(3*I-1)) 425,435,435
      425 IF (KK - 80) 430,430,410
      430 KK = 0
      435 KK = KK + KF(3*I-1) + KF(3*I-2)
          IF (KK - 80) 450,450,440
      440 WRITE (LINES,445)
      445 FORMAT (34HOSPECIFICATIONS EXCEED 80 COLUMNS.)
          CALL EXIT
      450 RETURN
          END

          SUBROUTINE MISS (M,NV,KARDS,LINES)
C
C         ROUTINE TO READ AND ECHO MISSING DATA SIGNAL CARD (MDSC)
C
C         COLUMNS OF MDSC CORRESPOND TO VARIABLES IN SERIAL ORDER OF INPUT.
```

```
C       A 1 IN A GIVEN COLUMN SIGNALS THAT ZERO SCORES FOR THE CORRES-
C       PONDING VARIABLE ARE TO BE EXCLUDED FROM THE ANALYSIS.
C       A 0 RESULTS IN INCLUSION OF ALL SCORES (FOR THAT VARIABLE) IN
C       THE ANALYSIS.
C
C       M = ONE-DIMENSIONAL VECTOR RETURNED CONTAINING NV ONES AND ZEROS.
C       NV = NUMBER OF VARIABLES.
C       KARDS = DEVICE NUMBER FOR CARD READER.
C       LINES = DEVICE NUMBER FOR LINE PRINTER.
C
C       THIS ROUTINE REFERENCES SUBROUTINE DCODE.
C
C       JAMES H. HOGGE, GEORGE PEABODY COLLEGE, NASHVILLE, TENNESSEE
C       VERSION OF 1/1/72
C
        DIMENSION KARD(80), M(1)
C
C       INPUT CARD.
C
        READ (KARDS,5) KARD
      5 FORMAT (80A1)
C
C       CONVERT SUCCESSIVE COLUMNS TO NUMBERS.
C
        DO 10 I = 1,NV
        CALL DCODE (KARD,I,I,M(I),IER)
C
C       CHECK FOR ERROR.
C
        IF (IER) 10,10,20
     10 CONTINUE
        WRITE (LINES,15) (M(I), I = 1,NV)
     15 FORMAT (27HOMISSING DATA SIGNAL CARD = // 1X, 80I1)
        RETURN
C
C       ERROR ROUTINE.
C
     20 WRITE (LINES,25) IER, KARD
     25 FORMAT (28HOINVALID CHARACTER IN COL., I3, 44H OF PRESUMED MISSING
       1 DATA SIGNAL CARD BELOW. / 1X, 80A1)
        CALL EXIT
        END

        SUBROUTINE MPRT (X,NR,NC,LINES,ND)
C
C       ROUTINE TO PRINT A MATRIX OR A VECTOR
C
C       X = ONE- OR TWO-DIMENSIONAL REAL ARRAY.
C       NR = NUMBER OF ROWS IN MATRIX X OR NUMBER OF ELEMENTS IN VECTOR X.
C       NC = NUMBER OF COLUMNS IN MATRIX X.  IF X IS A VECTOR, SET NC = 1.
C       LINES = OUTPUT DEVICE NUMBER.
C       ND = NUMBER OF ROWS DIMENSIONED FOR MATRIX X OR NUMBER OF ELEMENTS
C            DIMENSIONED FOR VECTOR X.
C
C       JAMES H. HOGGE, GEORGE PEABODY COLLEGE, NASHVILLE, TENNESSEE
C       VERSION OF 1/1/72
C
        DIMENSION X(1), Y(8)
C
C       CHECK TO SEE IF X IS A VECTOR.
C
        IF (NC - 1) 5,5,40
C
C       PRINT VECTOR.
C
      5 DO 30 I = 1,NR,8
        IF (I + 7 - NR) 10,15,15
     10 J = I + 7
        GO TO 20
     15 J = NR
     20 WRITE (LINES,25) (K, K = I,J)
     25 FORMAT (/ 2X, 8I9)
     30 WRITE (LINES,35) (X(K), K = I,J)
     35 FORMAT (5X, 8F9.3)
        RETURN
C
C       PRINT MATRIX.
C
     40 DO 65 I = 1,NC,8
        IF (I + 7 - NC) 45,50,50
     45 J = I + 7
        GO TO 55
     50 J = NC
     55 WRITE (LINES,25) (K, K = I,J)
        DO 65 L = 1,NR
```

```
      DO 60 M = I,J
      KK = L + ND * (M - 1)
      LL = M - I + 1
   60 Y(LL) = X(KK)
   65 WRITE (LINES,7C) L, (Y(N), N = 1,LL)
   70 FORMAT (I5, 8F9.3)
      RETURN
      END

      SUBROUTINE MPRTN (X,N,ND,LINES,NFILE)
C
C     ROUTINE TO PRINT A SQUARE MATRIX OR A VECTOR WITH VARIABLE NAMES
C
C     X = ONE- OR TWO-DIMENSIONAL REAL ARRAY.
C     N = NUMBER OF ROWS AND COLUMNS IN MATRIX X OR NUMBER OF ELEMENTS
C         IN VECTOR X.
C     ND = NUMBER OF ROWS DIMENSIONED FOR MATRIX X.  IF X IS A VECTOR,
C         SET ND = 1.
C     LINES = OUTPUT DEVICE NUMBER
C     NFILE = NUMBER OF DISK FILE IN WHICH VARIABLE NAMES ARE STORED,
C             20 CHARACTERS PER NAME, ONE NAME PER RECORD, ONE
C             CHARACTER PER WORD.
C
C     JAMES H. HOGGE, GEORGE PEABODY COLLEGE, NASHVILLE, TENNESSEE
C     VERSION OF 1/1/72
C
      DIMENSION X(1), Y(6), NAME(20)
C     CHECK TO SEE IF X IS A VECTOR.
      IF (ND - 1) 5,5,20
C
C     PRINT VECTOR.
C
    5 DO 10 I = 1,N
      READ (NFILE'I) NAME
   10 WRITE (LINES,15) I, NAME, X(I)
   15 FORMAT (I5, 1X, 20A1, F12.3)
      RETURN
C
C     PRINT MATRIX.
C
   20 DO 50 I = 1,N,6
      IF (I + 5 - N) 25,30,30
   25 J = I + 5
      GO TO 35
   30 J = N
   35 WRITE (LINES,40) (K, K = I,J)
   40 FORMAT (23X, 6I9)
      DO 50 L = 1,N
      READ (NFILE'L) NAME
      DO 45 M = I,J
      KK = L + ND * (M - 1)
      LL = M - I + 1
   45 Y(LL) = X(KK)
   50 WRITE (LINES,55) L, NAME, (Y(K), K = 1,LL)
   55 FORMAT (I5, 1X, 20A1, 6F9.3)
      RETURN
      END

      SUBROUTINE ROUND (X,N,Y)
C
C     ROUTINE TO ROUND TO N DECIMAL PLACES
C
C     X = VALUE TO BE ROUNDED.
C     N = NUMBER OF DECIMAL PLACES TO WHICH TO ROUND.
C     Y = ROUNDED VALUE (RETURNED).
C
C     X MAY BE NEGATIVE.
C
C     JAMES H. HOGGE, GEORGE PEABODY COLLEGE, NASHVILLE, TENNESSEE
C     VERSION OF 1/1/72
C
      IF (X) 5,15,10
    5 Y = X - 5.0 * 10.0 ** (-N-1)
      RETURN
   10 Y = X + 5.0 * 10.0 ** (-N-1)
      RETURN
   15 Y = X
      RETURN
      END

      SUBROUTINE START (KP,KARDS,LINES)
C
C     ROUTINE TO READ AND PRINT TITLE CARD(S) AND READ AND DECODE
C         PARAMETER CARD
C
```

```
C     KP.= ONE-DIMENSIONAL, 20-ELEMENT INTEGER ARRAY RETURNED CONTAINING
C          ENTRIES IN 4-COLUMN FIELDS OF PARAMETER CARD.
C     KARDS = DEVICE NUMBER FOR CARD READER.
C     LINES = DEVICE NUMBER FOR LINE PRINTER.
C
C     TITLE CARD(S) PRECEDE PARAMETER CARD AND MUST HAVE NON-NUMERIC
C          CHARACTER PUNCHED IN COLUMN ONE.
C
C     THIS ROUTINE REFERENCES SUBROUTINE DCODE.
C
C     JAMES H. HOGGE, GEORGE PEABODY COLLEGE, NASHVILLE, TENNESSEE
C     VERSION OF 1/1/72
C
      DIMENSION KK(80), KP(20)
      DATA KBLNK /' '/
C
C     READ CARD AND CHECK TO SEE IF IT IS BLANK.
C
      READ (KARDS,5) KK
    5 FORMAT (80A1)
      DO 10 I = 1,80
      IF (KK(I) - KBLNK) 25,10,25
   10 CONTINUE
      WRITE (LINES,15)
   15 FORMAT (38HOEND OF JOB -- BLANK CARD ENCOUNTERED.)
      CALL EXIT
C
C     READ SUBSEQUENT CARD(S).
C
   20 READ (KARDS,5) KK
C
C     CHECK FOR BLANK OR NUMERAL IN COLUMN ONE.
C
   25 CALL DCODE (KK,1,1,N,IER)
      IF (IER) 40,40,30
C
C     PRINT TITLE CARD.
C
   30 WRITE (LINES,35) KK
   35 FORMAT (1X, 80A1)
      GO TO 20
C
C     DECODE ENTRIES OF PARAMETER CARD.
C
   40 DO 55 I = 1,20
      CALL DCODE (KK,(I-1)*4+1,(I-1)*4+4,KP(I),IER)
      IF (IER) 55,55,45
   45 WRITE (LINES,50) IER, KK
   50 FORMAT (28HOINVALID CHARACTER IN COLUMN, I3, 41H OF SUPPOSED PARAM
     1ETER CARD ON NEXT LINE. / 1X, 80A1)
      CALL EXIT
   55 CONTINUE
      RETURN
      END

      SUBROUTINE TAKE (KF,X,NV,KARDS,LINES)
C
C     ROUTINE TO INPUT DATA CARDS AND EXTRACT SCORES ACCORDING TO FORMAT
C     SPECIFICATIONS STORED IN ARRAY BY SUBROUTINE FMAT.
C
C     KF = ONE-DIMENSIONAL INTEGER ARRAY IN WHICH FORMAT SPECIFICATIONS
C          HAVE BEEN STORED BY SUBROUTINE FMAT.
C     X = ONE-DIMENSIONAL REAL ARRAY IN WHICH SCORES ARE TO BE STORED.
C     NV = NUMBER OF SCORES TO BE TAKEN FROM DATA CARDS AND STORED IN X.
C     KARDS = DEVICE NUMBER FOR CARD READER.
C     LINES = DEVICE NUMBER FOR LINE PRINTER.
C
C     JAMES H. HOGGE, GEORGE PEABODY COLLEGE, NASHVILLE, TENNESSEE
C     VERSION OF 1/1/72
C
      DIMENSION KF(1), X(1), DIGTS(20), NUMS(10), KK(80)
      DATA KBLNK, MINUS, KDECI /' ','-','.'/
      DATA NUMS /'0','1','2','3','4','5','6','7','8','9'/
      NVF = 0
    5 NF = 1
C
C     READ A DATA CARD.
C
   10 READ (KARDS,15) KK
   15 FORMAT (80A1)
      NC = 1
C
C     CHECK FOR SKIP TO NEXT CARD OR END OF FORMAT SPECIFICATIONS.
C
      IF (KF(3*NF-1)) 20,125,25
```

```
   20 NF = NF + 1
      GC TC 10
C
C     EXTRACT SCCRE.
C
   25 NVF = NVF + 1
      X(NVF) = C.O
      NEGF = C
      KCGF = O
      NCGTS = C
      KCECF = C
      NC = NC + KF(3*NF-2)
      NE = NC + KF(3*NF-1) - 1
      IF (NE - 8C) 3C,30,150
   3C CC 90 I = NC,NE
      IF (KK(I) - KBLNK) 40,35,40
   35 J = 1
      GC TC 80
   4C CC 50 J = 1,1C
      IF (KK(I) - NUMS(J)) 50,45,5C
   45 KCGF = 1
      GC TC 80
   50 CCNTINUE
      IF (KK(I) - MINUS) 65,55,65
   55 IF (KCGF) 60,60,155
   60 NEGF = 1
      GC TC 90
   65 IF (KK(I) - KCECI) 155,70,155
   7C IF (KCECF) 75,75,155
   75 KCECF = NCGTS + 1
      GC TC 90
   8C NCGTS = NCGTS + 1
      IF (NCGTS - 20) 85,85,155
   85 CIGTS(NCGTS) = J - 1
   9C CCNTINUE
      NC = NE + 1
      CC 95 I = 1,NCGTS
   95 X(NVF) = X(NVF) + CIGTS(I) * 10.0 ** (NCGTS - I)
      IF (KCECF) 1C5,1C5,1CC
  100 X(NVF) = X(NVF) * 10.C ** (KCECF - NCGTS - 1)
      GC TC 115
  105 IF (KF(3*NF)) 115,115,11C
  11C X(NVF) = X(NVF) * 1C.C ** (-KF(3*NF))
  115 IF (NEGF) 125,125,120
  120 X(NVF) = -X(NVF)
  125 IF (NVF - NV) 130,135,135
  13C NF = NF + 1
      IF (KF(3*NF-1)) 20,5,25
  135 NF = NF + 1
      IF (KF(3*NF-1)) 145,14C,14C
  14C RETLRN
C
C     SKIP ACCITICNAL CARCS AS RECLIRED.
C
  145 READ (KARCS,15) KK(1)
      GC TC 135
C
C     ERRCR RCLTINE.
C
  150 I = 81
  155 WRITE (LINES,160) I, KK
  16C FCRMAT (15HOERRCR NEAR CCL, I3, 19H CF DATA CARD BELCW / 1X, 8CA1)
      CALL EXIT
      ENC

      SLBROUTINE TAKEC (KF,X,NV,LINES)
C
C     VERSION CF TAKE USING SUBRCUTINE REAC FROM IBM CSP
C
C     RCUTINE TC INPLT CATA CARCS AND EXTRACT SCCRES ACCCRCING TC FORMAT
C     SPECIFICATICNS STCREC IN ARRAY BY SUBRCUTINE FMAT.
C
C     KF = ONE-CIMENSICNAL INTEGER ARRAY IN WHICF FORMAT SPECIFICATIONS
C          HAVE BEEN STCREC BY SLBRCUTINE FMAT.
C     X = ONE-CIMENSIONAL REAL ARRAY IN WHICH SCORES ARE TC BE STORED.
C     NV = NUMBER OF SCCRES TC BE TAKEN FRCM DATA CARDS AND STORED IN X.
C     LINES = CEVICE NLMBER FCR LINE PRINTER.
C
C     JAMES H. HCGGE, GECRGE PEABCCY CCLLEGE, NASHVILLE, TENNESSEE
C     VERSION CF 1/1/72
C
      CIMENSICN KF(1), X(1), CIGTS(20), NUMS(1C), KK(8C)
      CATA KBLNK, MINUS, KCECI /' ','-','.'/
      CATA NUMS /'0','1','2','3','4','5','6','7','8','9'/
      NVF = 0
    5 NF = 1
```

```
C
C      READ A DATA CARD.
C
   1C CALL READ (KK,1,80,NER)
      NC = 1
C
C      CHECK FOR SKIP TC NEXT CARC CR END CF FORMAT SPECIFICATIONS.
C
      IF (KF(3*NF-1)) 15,120,20
   15 NF = NF + 1
      GC TC 10
C
C      EXTRACT SCORE.
C
   20 NVF = NVF + 1
      X(NVF) = C.C
      NEGF = 0
      KCGF = 0
      NCGTS = C
      KCECF = C
      NC = NC + KF(3*NF-2)
      NE = NC + KF(3*NF-1) - 1
      IF (NE - 80) 25,25,145
   25 CC 85 I = NC,NE
      IF (KK(I) - KBLNK) 35,30,35
   3C J = 1
      GC TC 75
   35 CC 45 J = 1,1C
      IF (KK(I) - NUMS(J)) 45,4C,45
   4C KCGF = 1
      GC TC 75
   45 CONTINUE
      IF (KK(I) - MINUS) 60,50,6C
   5C IF (KCGF) 55,55,15C
   55 NEGF = 1
      GC TO 85
   60 IF (KK(I) - KCECI) 150,65,150
   65 IF (KCECF) 70,70,150
   70 KCECF = NCGTS + 1
      GC TO 85
   75 NCGTS = NCGTS + 1
      IF (NCGTS - 20) 80,80,15C
   8C CICTS(NCCTS) = J - 1
   85 CONTINUE
      NC = NE + 1
      CC 90 I = 1,NCGTS
   9C X(NVF) = X(NVF) + CIGTS(I) * 10.0 ** (NCGTS - I)
      IF (KDECF) 1C0,1C0,95
   95 X(NVF) = X(NVF) * 10.0 ** (KCECF - NCGTS - 1)
      GC TO 11C
  1CC IF (KF(3*NF)) 110,110,1C5
  1C5 X(NVF) = X(NVF) * 10.0 ** (-KF(3*NF))
  110 IF (NEGF) 120,120,115
  115 X(NVF) = -X(NVF)
  120 IF (NVF - NV) 125,13C,13C
  125 NF = NF + 1
      IF (KF(3*NF-1)) 15,5,2C
  130 NF = NF + 1
      IF (KF(3*NF-1)) 14C,135,135
  135 RETURN
C
C      SKIP ACCITICNAL CARDS AS RECLIRED.
C
  140 CALL REAC (KK,1,1,NER)
      GC TC 130
C
C      ERRCR RCLTINE.
C
  145 I = 81
  150 WRITE (LINES,155) I, KK
  155 FCRMAT (15HCERRCR NEAR CCL, I3, 19H CF DATA CARD BELCW / 1X, 80A1)
      CALL EXIT
      ENC

      SUBROUTINE TAKEI (KF,X,NV,ID,NL,NR,KARDS,LINES)
C
C      ROUTINE TC INPUT DATA CARDS, RETURNING A SUBJECT IDENTIFICATION
C      FIELD AND SCORES EXTRACTED ACCORDING TO FORMAT SPECIFICATIONS
C      STORED IN ARRAY BY SUBROUTINE FMAT
C
C      KF = ONE-CIMENSIONAL INTEGER ARRAY IN WHICH FORMAT SPECIFICATIONS
C           HAVE BEEN STORED BY SUBROUTINE FMAT.
C      X = ONE-DIMENSIONAL REAL ARRAY IN WHICH SCORES ARE TO BE STORED.
C      NV = NUMBER OF SCORES TO BE TAKEN FROM DATA CARDS AND STORED IN X.
C      ID = ONE-CIMENSIONAL INTEGER ARRAY RETURNED CONTAINING THE
```

```
C           IDENTIFICATION FIELD, ONE ALPHAMERIC CHARACTER PER ELEMENT.
C        NL = LEFT-MOST COLUMN OF SUBJECT IDENTIFICATION FIELD.
C        NR = RIGHT-MOST COLUMN OF SUBJECT IDENTIFICATION FIELD.
C        KARDS = DEVICE NUMBER FOR CARD READER.
C        LINES = DEVICE NUMBER FOR LINE PRINTER.
C
C        THE SUBJECT IDENTIFICATION FIELD MUST APPEAR ON THE FIRST CARD OF
C           THE SUBJECT'S CARD SET.
C
C        JAMES H. HOGGE, GEORGE PEABODY COLLEGE, NASHVILLE, TENNESSEE
C        VERSION OF 1/1/72
C
         DIMENSION KF(1), X(1), DIGTS(20), NUMS(10), KK(80), ID(1)
         DATA KBLNK, MINUS, KDECI /' ','-','.'/
         DATA NUMS /'0','1','2','3','4','5','6','7','8','9'/
         NCR = 0
         NE = 0
         NVF = 0
       5 NF = 1
C
C        READ A DATA CARD.
C
      10 READ (KARDS,15) KK
      15 FORMAT (80A1)
         NCR = NCR + 1
C
C        CHECK TO SEE IF THIS IS FIRST CARD.
C
         IF (NCR - 1) 30,20,30
C
C        COPY ID FIELD INTO ID.
C
      20 DO 25 I = NL,NR
         NE = NE + 1
      25 ID(NE) = KK(I)
      30 NC = 1
C
C        CHECK FOR SKIP TO NEXT CARD OR END OF FORMAT SPECIFICATIONS.
C
         IF (KF(3*NF-1)) 35,140,40
      35 NF = NF + 1
         GO TO 10
C
C        EXTRACT SCORE.
C
      40 NVF = NVF + 1
         X(NVF) = 0.0
         NEGF = 0
         KDGF = 0
         NDGTS = 0
         KDECF = 0
         NC = NC + KF(3*NF-2)
         NE = NC + KF(3*NF-1) - 1
         IF (NE - 80) 45,45,165
      45 DO 105 I = NC,NE
         IF (KK(I) - KBLNK) 55,50,55
      50 J = 1
         GO TO 95
      55 DO 65 J = 1,10
         IF (KK(I) - NUMS(J)) 65,60,65
      60 KDGF = 1
         GO TO 95
      65 CONTINUE
         IF (KK(I) - MINUS) 80,70,80
      70 IF (KDGF) 75,75,170
      75 NEGF = 1
         GO TO 105
      80 IF (KK(I) - KDECI) 170,85,170
      85 IF (KDECF) 90,90,170
      90 KDECF = NDGTS + 1
         GO TO 105
      95 NDGTS = NDGTS + 1
         IF (NDGTS - 20) 100,100,170
     100 DIGTS(NDGTS) = J - 1
     105 CONTINUE
         NC = NE + 1
         DO 110 I = 1,NDGTS
     110 X(NVF) = X(NVF) + DIGTS(I) * 10.0 ** (NDGTS - I)
         IF (KDECF) 120,120,115
     115 X(NVF) = X(NVF) * 10.0 ** (KDECF - NDGTS - 1)
         GO TO 130
     120 IF (KF(3*NF)) 130,130,125
     125 X(NVF) = X(NVF) * 10.0 ** (-KF(3*NF))
     130 IF (NEGF) 140,140,135
     135 X(NVF) = -X(NVF)
     140 IF (NVF - NV) 145,150,150
```

```
      145 NF = NF + 1
          IF (KF(3*NF-1)) 35,5,40
      150 NF = NF + 1
          IF (KF(3*NF-1)) 160,155,155
      155 RETURN
C
C         SKIP ADDITIONAL CARDS AS REQUIRED.
C
      160 READ (KARDS,15) KK(1)
          GO TO 150
C
C         ERROR ROUTINE.
C
      165 I = 81
      170 WRITE (LINES,175) I, KK
      175 FORMAT (15HOERROR NEAR CCL, I3, 19H OF DATA CARD BELOW / 1X, 80A1)
          CALL EXIT
          END
```

Appendix B:
Peabody Computer Center User's Manual*

CONTENTS

1. Introduction

2. Policies

   2.1 Computer Center Regulations
   2.2 Establishing an Account
   2.3 Extent of Services

3. Utilizing the IBM 1130

   3.1 IBM 1130 System Control Cards
   3.2 The Peabody Time-Accounting System
   3.3 Computer Operations
   3.4 Application Software Available

4. Unit Record Operations

   4.1 Data Card Preparation
   4.2 Tab Card Preparation
   4.3 Card Punch Operating Features

5. Courses Offered Relating to Computing and Data Processing

6. Appendix

   6.1 FORTRAN Programming References

*This manual was prepared by the staff of the George Peabody College Computer Center.

1. INTRODUCTION

The George Peabody College Computer Center is a campuswide facility established primarily for the support of instruction and research in the University Center. The operation of the Center is underwritten by funds from the College and the John F. Kennedy Center for Research on Education and Human Development.

The Computer Center is located in the Human Development Laboratory on the ground floor, and currently houses an IBM 1130 computer. Equipment available in support of work on the computer includes six card punch machines, a card verifier, a card reproducer, and a counter-sorter with pocket counters.

2. POLICIES

2.1 Computer Center Regulations

The Computer Center is open to all students, researchers, faculty and administrators who have need for computing and data processing services. The Center is operated under an open shop arrangement and users are expected to show respect for the equipment and rights of the other users by observing a minimum of safety rules and operation regulations.

In order that the Computer Center may serve users in the most efficient manner, users are requested to observe the following regulations:

Punch Cards

1. Remove used cards from all machines, e.g., key punch, and 514 reproducer.
2. Do not use top of computing equipment as work areas.
3. Replace library program cards in the file cabinet.
4. Place used cards in wastebaskets.
5. Turn off unit record equipment after use.

Paper

1. Remove program printout and results from printer after program is completed and/or after a series of jobs.
2. Place unwanted printouts in wastebaskets.

Housekeeping

Since humidity, temperature, and dust control are critical factors in the operation of equipment, it is very important that users:

1. Refrain from smoking in the Center.
2. Refrain from eating and drinking in the Center.
3. Use the wastebaskets provided for discarding paper, cards, etc.

Key-Holders

1. The entrance to the Center should be locked while the key-holder is working after hours.

2. Key-holders are requested to notify computer center staff if they plan to work in the center after regularly scheduled hours.
3. No unauthorized users should use the equipment while the key-holder is working after hours.
4. Key-holders may not loan their keys to other users.
5. Key-holders should clean up, turn off equipment and lights, and lock all doors before leaving the Center.

Telephone

The telephone located in the Computer Center is primarily for use by Computer Center staff. Incoming calls for users should be limited to emergency use only.

2.2 Establishing an Account

Application for a Computer Center account is made on a form available at the Computer Center. Although account numbers will be assigned to students in a course utilizing the computer, the instructor (rather than the students) should make application for time for students in his course.

Application may also be made for computer time for unsponsored activities. Forms for this purpose are available at the Computer Center.

An account number consists of five digits. The first digit indicates whether the project is sponsored or unsponsored (1 = unsponsored, 2 = sponsored); the second, the category of use. The last three digits designate the number of the individual project within the support status and category of use combination.

The categories of use are the following:

Second Digit

0	Development: This includes all "in house" software development.
1	Administrative data processing.
2	On-campus research by faculty, staff, and students.
3	Off-campus research by faculty, staff, and students.
4	Other off-campus accounts.
5	Computer Center contracts.
6	IBM maintenance.
7	Housekeeping (i.e,, system upkeep; billing).
8	Instruction.
9	Instruction.

Sponsored accounts are currently charged for 1130 time at the rate of $50 per hour of actual CPU (core) time. There is no charge for use of other Computer Center equipment.

2.3 Extent of Services

Cards and Paper

Stock forms (computer paper) are furnished for all users by the Computer Center; see the Computer Center staff concerning availability of cards.

Card Punching

The Computer Center does not offer keypunch services. Instead, the user requiring card punching should provide his own keypunch operator. Also, keypunch services are available from various commercial data processing firms in Nashville.

Card Storage

Users may apply for drawer storage using forms available at the Computer Center. Limited active storage space will be assigned on a first-come, first-served basis.

Computing and Data Processing Consultation

The staff of the Computer Center offers assistance to the user in debugging his FORTRAN program or learning about the software available through the Center. Assistance in debugging one's program or utilizing the Peabody Statistical Library may be obtained from anyone "on duty" in HDL 105 or by calling extension 8031.

Machine Operation

The Computer Center user is expected to operate the machines of the Center with supervisory assistance to the extent needed. Assistance in learning to operate the machines may be obtained from the Computer Center staff.

Instruction

A major portion of the Center's efforts is devoted to instruction. Information concerning the instructional program of the Center may be obtained from any staff member.

Programming

One of the primary responsibilities of the Computer Center staff is the development and maintenance of software relevant to the instructional programs. Although the staff will assist users in the writing and debugging of their own computer programs, the Center staff cannot do programming of a specific nature for individual users or projects. Instead, the Center maintains a list of individuals who are available for special programming by the hour or job.

3. UTILIZING THE IBM 1130

3.1 IBM 1130 System Control Cards

The following comments apply only to FORTRAN jobs; furthermore, only basic control cards are listed. The user desiring information concerning other system options should see the IBM manual *1130 Disk Monitor System, Version 2, Programming and Operator's Guide*. This publication is included in the manual rack in HDL 107.

CONTROL CARDS FOR MAINLINE PROGRAM

1. // JOB T
2. // XEQ ON
3. Account card
4. // FOR
5. *ONE WORD INTEGERS
6. *LIST SOURCE PROGRAM
7. IOCS card
8. FORTRAN Statements
9. // XEQ
10. Data cards (if any)
11. // XEQ OFF
12. Blank card

CONTROL CARDS FOR MAINLINE PROGRAM:
CARD PREPARATION

1. // JOB T.   This card is punched exactly as follows:

Column	Entry
1–2	//
3	Blank
4–6	JOB
7	Blank
8	T
9–80	Blank

2. // XEQ ON.   This card is punched exactly as follows:

Column	Entry
1–2	//
3	Blank
4–6	XEQ
7	Blank
8–9	ON
10–80	Blank

3. Account Card.　This card is punched as follows:

Column	Entry
1	(period)
2–6	Account number
7–10	Blank
11–25	Name, job description, etc. Any desired combination of keypunch characters may be supplied. This information will be printed verbatim on the output and will be entered in the job log.
26–30	Blank
31–34	Code word (see section 3.2).
35–80	Blank

4. // FOR.　This card is punched exactly as follows:

Column	Entry
1–2	//
3	Blank
4–6	FOR
7–80	Blank

5. *ONE WORD INTEGERS.　This card is punched exactly as follows:

Column	Entry
1	*
2–4	ONE
5	Blank
6–9	WORD
10	Blank
11–18	INTEGERS
19–80	Blank

6. *LIST SOURCE PROGRAM.　This card is punched exactly as follows:

Column	Entry
1	*
2–5	LIST

6	Blank
7–12	SOURCE
13	Blank
14–20	PROGRAM
21–80	Blank

7. *IOCS Card. This card names the input-output devices the program (or any of its subprograms) will use. The card always begins as follows:

Column	Entry
1	*
2–5	IOCS
6	(

The left parenthesis is followed by device names, listed exactly as shown in the following table:

I/O Device	IOCS Designation
1442 card READ/PUNCH	CARD
1403 printer	1403PRINTER
Either disk drive	DISK
Console printer	TYPEWRITER
Console keyboard	KEYBOARD
Unformatted disk I/O (tape statements)	UDISK

Device names are separated by commas, and the last device name is followed by a right parenthesis.

*IOCS(CARD,1403PRINTER)

would specify that the program will use the 1442 card READ/PUNCH and the 1403 printer.

8. FORTRAN Statements. For a discussion of 1130 FORTRAN, see the IBM manual *IBM 1130/1800 Basic FORTRAN IV Language*. The CALL EXIT statement should be used to terminate the program (rather than the STOP statement, which results in a pause requiring operator intervention).

9. // XEQ. This card is punched exactly as follows:

Column	Entry
1–2	//
3	Blank
4–6	XEQ
7–80	Blank

10. Data Cards (if any). The layout of these cards will, of course, vary from program to program.

11. // XEQ OFF.

Column	Entry
1–2	//
3	Blank
4–6	XEQ
7	Blank
8–10	OFF
11–80	Blank

12. Blank Card. This card may be omitted if this job is to be included among other jobs in a "stacked job environment."

CONTROL CARDS FOR MAINLINE PROGRAM WITH ONE OR MORE SUBPROGRAMS

Card Order

1. // JOB T
2. // XEQ ON
3. Account card
4. // FOR
5. *ONE-WORD INTEGERS
6. *LIST SOURCE PROGRAM
7. Subprogram FORTRAN statements
8. // DUP
9. *STORE card
10. // FOR
11. *ONE WORD INTEGERS
12. *LIST SOURCE PROGRAM
13. *IOCS card
14. Mainline FORTRAN statements
15. // XEQ
16. Data cards (if any)
17. // XEQ OFF
18. Blank card

Items 4 through 9 are repeated for each subprogram. The mainline program and its system control cards must appear last in the input sequence, since the 1130 requires all subprograms to have been stored on disk at the time of mainline execution.

CONTROL CARDS FOR MAINLINE PROGRAM WITH ONE OR MORE SUBPROGRAMS

Card Preparation

Note: Cards already described above [cards 1 through 7] are not listed below.

8. // DUP. This card is punched exactly as follows:

Column	Entry
1–2	//
3	Blank
4–6	DUP
7–80	Blank

9. *STORE Card. This card is punched as follows:

Column	Entry
1	*
2–6	STORE
7–12	Blank
13–14	WS
15–16	Blank
17–18	UA
19–20	Blank
21–25	Subprogram name, beginning in column 21
26–80	Blank

3.2 The Peabody Time-Accounting System

The IBM 1130 has no internal time clock; hence, this computer is unable to keep track of user job times without input of external clock readings. Fortunately, the 1130 does feature a console clock which runs only when the central processing unit (the "core" or "brain" of the 1130) is actually in operation. Although the 1130 cannot sense the reading of the console clock directly, the computer can be programmed to accept user entry of console clock readings.

As described in Section 3.1, above, each job begins with the execution of ON, the sign-on program. [Program] ON reads the account card, requests entry of the current console clock reading (via the console keyboard), and notes the user's arrival. Each job ends with the execution of OFF, the sign-off program. [Program] OFF requests entry of the current clock reading and enters the user's account number, account card information and time consumption in the job log.

The code word is an important feature of the system, and all accounts are set up with a code word. A utility program is provided for the purpose of changing one's account code word. This program, U01, is stored on disk and requires a single control card in addition to a // XEQ U01 card. The control card contains the following information:

Column	Entry
1–5	Account number
6–10	Blank
11–14	Current code word

15–20	Blank
21–24	New code word
25–80	Blank

The code word may consist of any combination of blanks, letters, or numerals.

Let us suppose that Sam Jones has only four blanks for the code word of his account number 22022 and wishes to specify LB73 as his code word. The complete card setup to enter his code word would consist of the following sequence (where b = blank):

```
//bJOBbT
//bXEQbON
.22022bbbbSAMbJONES
//bXEQbU01
22022bbbbbbbbbbbbbbbbLB73
//bXEQbOFF
Blank Card
```

Please note that this job will run with an account card on which the code word field is blank. After this job has been run, Sam must supply his new code word on the account card for subsequent jobs.

As a second example, consider Betty Smith (account number 11011) who wishes to change her code word from BETS to BESM. She would run the following job:

```
//bJOBbT
//bXEQbON
.11011bbbbBETTYbSMITHbbbbbbbbbbBETS
//bXEQbU01
11011bbbbbBETSbbbbbbBESM
//bXEQbOFF
Blank Card
```

The successful execution of U01 always results in the printing of an acknowledgement on the 1403 printer, so be sure to check your output.

The successful execution of OFF is dependent upon the successful execution of the job it follows. If an error occurs in the compilation or execution of any routine in the job, the execution of OFF will be suppressed. This situation will result in no additional charges to the user if the next job in the input stream begins with the successful execution of ON, which is normally the case. If, however, any subsequent core activity occurs without the prior execution of ON, this activity will be charged to the user who did not successfully execute OFF. In cases where the user wishes to avoid the possibility of undesired charges, the following "dummy job" may be run in order to make certain sign-off has occurred.

```
// JOB T
// XEQ OFF
Blank Card
```

Users desiring to utilize a disk configuration which will result in the system cartridge being removed or disabled must execute ON before altering the cartridge configuration for their job. After restoring the system to the standard cartridge configuration, the same user should run the dummy job described above to sign off.

Questions regarding the Peabody Time-Accounting System may be directed to Computer Center staff, who will be happy to assist you in every way possible.

3.3 Computer Operations

The Computer Center user is expected to operate the equipment of the Center with a minimum of supervisory assistance. The following outline will assist users in normal operation.

1. Press NPRO on card reader and remove any cards from stacker.
2. Place cards in hopper of reader and press START.
3. Ready printer by pressing START on printer.
4. Press PROGRAM START on console.
5. Enter beginning console clock time upon request.
6. Enter ending console clock time upon request.
7. Press NPRO (if no additional program follows yours).
8. Remove all cards from stacker.
9. Remove results from printer by pressing STOP and CARRIAGE RESTORE *only* if no cards are still in the input hopper of the reader.

Note: In case of card jam or suspected equipment failure, consult a staff member.

Users interested in more details of equipment operation should consult *IBM 1130 Operating Procedures Form A26-5717-0.*

3.4 Application Software Available

Several application software packages are operational on the 1130 computer. The following list is a partial list of application packages now avaiable to center users:

1. IBM scientific subroutine package (SSP)
2. IBM commercial subroutine package (CSP)
3. Peabody Statistical Library (PSL)
4. Project control system (PCS)
5. Information retrieval system (IRS)

The IBM scientific subroutine package (SSP) is a collection of I/Q-free FORTRAN subroutines for statistical analysis, matrix algebra, and certain mathematical techniques. The IBM commercial subroutine package (CSP) provides business-oriented capabilities in a FORTRAN environment. The SSP and CSP require the user to be familiar with FORTRAN. A manual for the SSP package is available from the Peabody Bookstore, and a manual for the CSP is available from the IBM Corporation.

The IBM statistical system and the Peabody Statistical Library assume no knowledge of FORTRAN, and may be utilized by the relatively inexperienced computer user. The IBM statistical system includes stepwise multiple linear regression, analysis of variance, factor analysis, and polynomial curve fitting. A manual is available from IBM. The Peabody Statistical Library (PSL) includes the following programs:

Descriptive statistics
Contingency table construction
Frequency tabulation for single-column variables
Frequency tabulation for two-column variables
Covariance and correlation
Correlations for incomplete data
Partial correlation
Spearman rank difference correlation
Kendall rank correlation
Bivariate scatter plots
Single-classification analysis of variance with multiple groups and/or trials
Double- or triple-classification analysis of variance
Analysis of variance, three-factor design with repeated measures on one factor
Analysis of variance, three-factor design with repeated measures on two factors
Treatments-by-treatments-by-subjects analysis of variance
T-test for independent means and F-tests for homogeneity of independent variances
T-tests for related measures
Simple analysis of covariance
Orthogonal comparisons
Generalized analysis of variance
Item analysis and test scoring (two programs)
Chi-square analysis
Chi-square analysis of pre-compiled contingency tables
Regression analysis with generation and/or transformation of variables
Factor analysis with rotations
Multiple discriminant analysis
Program to change account code word
Card-listing program
Screening for invalid punches
Summary punch program
Multiple-field, one-pass sorting
Card reproduction with data shifting

A program (MANOVA) is also available for multivariate analysis of variance, of covariance and of regression which provides an exact solution with equal or unequal numbers of observations in the cells. This program is not part of the Peabody Statistical Library, and information concerning MANOVA may be obtained from the Computer Center staff.

The Project Control System (PCS) is designed to assist management in planning and supervising projects. The system processes data from networks planned in either precedence-diagramming or arrow-diagramming methods, and all routines are disk resident. The sequence of processing and system output is controlled by the control cards. Output reports include status listings, bar charts, and basic resource and cost summarization reports. An operations manual is available on a loan basis from the computer staff.

The Information Retrieval System (IRS) is a collection of generalized programs capable of performing sophisticated data manipulation. The system is composed of programs including a generalized load routine, a selectivity routine, a sort routine, and a report generation routine. The system will output specified information fields from selected records of a data file, in virtually any given order or form. Documentation for the system may be obtained on a loan basis from the computer staff.

4. UNIT RECORD OPERATIONS

4.1 Data Card Preparation

INTRODUCTION

Pictured in Figure B-1 is a standard tab card with the punches obtained using the IBM 029 keypunch machine. There are 80 columns in the tab card. Any one column can contain a single character such as a letter of the alphabet, a numeral, or a special character.

For each character the keypunch machine automatically prints at the top of the column and punches the appropriate holes for that column. The computer reads only the holes punched. The color, printed format,

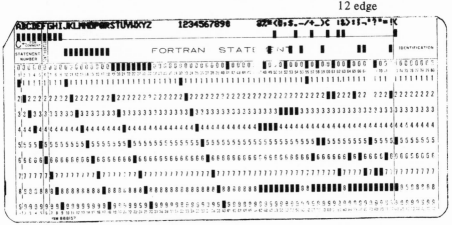

Figure B-1. Standard Tab Card Punched by IBM 029

and corner cuts of the card are for the user's benefit and are of no significance to the computer.

The printed side of the card is called the face of the card. The top of
the card is called the 12 edge; the bottom of the card is called the 9 edge.
Care should be taken to place the cards in the hopper of each machine
according to the directions on that particular machine.

4.2 Tab Card Preparation

1. Format. In order to make the most of the data collected, it is important that the data be organized and posted to cards in a manner compatible with the computer programs to be used. (See *Peabody Statistical
Library User's Manual*, Section 2, Preparation of Data for the Peabody
Statistical Library).

Each program requires a "data layout" of a particular form. It is
generally true that all variables for each subject are read at one time by
the computer. Data should be grouped by subjects rather than by variables.
Hence, data should be organized in the following general manner:

	Variables			
Subject	1	2	3	etc.
1	$X_{1,1}$	$X_{1,2}$	$X_{1,3}$	$\ldots$
2	$X_{2,1}$	$X_{2,2}$	$X_{2,3}$	$\ldots$
3	$X_{3,1}$	$X_{3,2}$	$X_{3,3}$	$\ldots$
etc.	$\ldots$	$\ldots$	$\ldots$	$\ldots$

Data are symbolized by X_{ij}, where i refers to the subject and j refers
to the variable.

It is suggested that the first 8 columns of a data card contain identification codes to identify the subject whose data follows on the card(s);
column nine, deck number, and column ten, card number. The codes contained in the first 8 columns could contain such information as sex,
social-economic-status, etc., and could be useful for sorting the cards in
order to form different groups for analysis using the computer. It is important to duplicate the identification field on each successive card of a
subject's set; this practice permits sorting while maintaining each subject's
card set intact.

2. Coding. Since the tab card has 80 columns, many users code data
to be punched on coding forms with the 80 columns marked off. There is
a one-to-one correspondence between the positions on a line of the coding
form and the columns on the card. Each line of the coding form represents
the information to be punched on a single card. Organizing data on coding
forms indicates exactly what belongs in a particular column. This is especially important if the user plans to have another person keypunch the
data. Keypunching services are usually charged on an hourly basis, and

posting data to coding forms tends to reduce the total time needed to punch the data on cards.

3. Keypunch Services. The Computer Center does not offer keypunching services; however, commercial services are available. It is advisable to require that data be verified by a second operator using the IBM 059 verifier machine. All data to be processed should be punched using the character set found on the IBM 029 keypunch machine. Consult the Computer Center staff if another type of keypunch machine is to be used to determine whether or not the punches will be compatible with the IBM 1130 computer.

4.3 Card Punch Operating Features

The Computer Center is currently using two different keypunch machines—models 026 and 029. The operations are essentially the same for both models. The following description of keypunch machines will help users to become familiar with either machine.

CARD PUNCH. The card punch can be used to punch numeric, alphabetic, and special-character information. The keyboard is very similar to the keyboard on a typewriter. A light touch on a key of the keyboard causes a punch or punches to be electrically driven through the card, thereby cutting one, two, or three holes in a column, according to the key used. As each column is punched, the card is automatically advanced to the next column.

CARD HOPPER. The card hopper, which holds approximately 500 cards, is located at the upper right of the machine. The cards are placed in the hopper, face forward, with the 9 edge down, and are fed front card first. A sliding pressure plate assures uniform feeding. A card is fed down from the hopper to the card bed by pressing a card-feed key.

PUNCHING STATION. Punching is performed at the first of two stations along the card bed. To start an operation, two cards are fed into the card bed at the right of the punching station. As the second card is fed, the first card is automatically registered for punching—that is, it is positioned at the punching station to the left. While the first card is being punched, the second card waits at the right of card bed. When column 80 of the first card passes the punching station, the card is transported to the next station, the second card is registered at the punching station, the next card in the hopper is fed down to the card bed. A single card can be placed in the card bed by hand and registered in punching position by pressing the registering key.

CARD STACKER. The card stacker is located at the upper left of the machine. After each card passes the reading station, it is fed into the stacker. Cards are stacked at an angle, 12 edge down.

BACKSPACE KEY. This key is located below the card between the reading and punching stations. As long as it is held down, the cards at the punching and reading stations are backspaced continuously. The program card, which controls skipping and duplicating, is also backspaced.

COLUMN INDICATOR. The column indicator is located at the base of the program drum holder and indicates the next column to be punched.

KEYBOARD. The keyboard is very similar to that of a typewriter. The letter keys are arranged according to the standard typewriter touch system. The digit keys are placed so that a rapid three-finger touch system can be used.

FEED KEY. This key causes a card in the card hopper to be fed to the right side of the card bed. This card must be registered for punching.

REGISTER KEY. This key positions a card in the right side of the card bed for punching. If a card is in the center of the card bed (between the reading and punching stations) it will be appropriately positioned for reading. Once a card is registered, it may not be released except by passing column 80 on the card or by pressing the release key.

RELEASE KEY. Depressing the release key causes a card being punched to skip the rest of the columns and to move to the center part of the card bed.

ALPHABETIC SHIFT KEY. Depressing this key causes the alphabetic character on a key to be punched.

NUMERIC SHIFT KEY. This key causes the numeric and special characters to be punched. The shift keys are effective only while depressed.

MULTIPLE-PUNCH KEY. If it is desired to put more than one punch into a column, depress the multiple-punch key before putting any punches in that column. Hold the key down until all the punches desired in the column have been made, then release it. The card will now advance to the next column.

DUPLICATE KEY. Depressing this key will cause the columns on the cards going through the reading station to be duplicated into the corresponding columns of the card being punched.

SKIP KEY. This key acts like the tab key on a typewriter and will cause a skip to the end of the field being punched when used in connection with the drum program card. When not using a drum program card, this key causes the same effect at the space bar.

PRINT SWITCH. When the print switch is turned on, the characters punched will also be printed on top of the card.

AUTOMATIC FEED SWITCH. If this switch is turned on, pressing the release key will be all that is necessary to release the card being punched, feed another card from the hopper and position it for punching. To start the automatic feed cycle, the feed key should be depressed twice. From then on, pressing the release key will be equivalent to pressing the release key, the feed key, and the register key in that order.

CLEAR SWITCH. When the card bed is to be completely cleared at the end of a punching operation without feeding more cards from the hopper, the automatic feed switch should be turned off during the punching of the next-to-the last card. The last card is then registered for punching by pressing the register key. If the automatic feed switch is turned off after the last card to be punched is completed, the release and register keys must be pressed alternately three times. Each user should clear the card bed after using the machine.

PROGRAM DRUM CARD. The program card controls automatic skipping, automatic duplicating, and shifting from numeric to alphabetic punching, or vice versa. Each of these operations is designated by a specific code punched in a program card.

Punch	*Function*
12 &(ampersand)	Defines the length of a field.
11 – (minus)	Starts automatic skipping.
0	Starts automatic duplication.
1	Shifts keyboard to numeric mode.
∅ (blank)	Indicates a field to be manually punched.
A	Continues alphabetic information.
&	Continues any other field.

The program card is mounted on the program drum for insertion in the machine. The starwheels are engaged for reading of the character.

5.0 COURSES OFFERED RELATING TO COMPUTING AND DATA PROCESSING

DEPARTMENT	COURSE NO.	TITLE	HOURS CREDIT
Chemistry	232	Instrumental Analysis	4
Education/Psych.	214	Statistical Analysis by Computer	2–3
Education	287	Automated Data Processing in Education	3
Education	390	Methods of Educational Research	3
Library Science	313	Systems Analysis in Library and Information Science	3
Library Science	381	Data Processing of Library Operations	3
Library Science	382	Documentation and Information Storage and Retrieval	3
Mathematics	201	Introduction to Numerical Analysis I	3
Mathematics	202	Introduction to Numerical Analysis II	3
Mathematics	235W	The Computer in the Secondary School	3
Physics	104	Introduction to Computer Applications	4
Physics	204	Computer Applications in the Sciences	4

Psychology	210B	Statistical Analysis	3
Psychology	211	Computer Applications in the Behavioral Sciences	3
Psychology	313	Factor Analysis	3
Psychology	319	Advanced Seminar in Statistics	1–3
Psychology	329	Advanced Seminar in Measurement	1–3
Psychology	320	Measurement and Correlation	4

6. APPENDIX

6.1 FORTRAN Programming References

Anderson, Decima M., *Computer Programming FORTRAN IV*. New York: Appleton-Century-Crofts, 1966.

Dimitry, D., and Mott, T., Jr., *Introduction to FORTRAN IV Programming*. New York: Holt, Rinehart & Winston, 1966.

Farina, M. V., *FORTRAN IV Self-Taught*. Englewood Cliffs, N.J.: Prentice-Hall, 1966.

Golden, J. R., *FORTRAN IV Programming and Computing*. Englewood N.J.: Prentice-Hall, 1965.

Hughes, Joan K., *Programming the IBM 1130*. New York: Wiley, 1969.

IBM Systems Reference Library:
  IBM 1130/1800 Basic FORTRAN IV Language.
  IBM 1130 DISK Monitor System, Version 2, Programming and Operators Guide.
  (See Manual Rack in Computer Room for additional IBM publications.)

IBM Project Control System, Version 2, Program Description Manual.

IBM Project Control System (1130-CP-05X) Operator's Manual.

Louden, R. K., *Programming the IBM 1130 and 1800*. Englewood Cliffs, N.J.: Prentice-Hall, 1967.

Mann, Richard A., *An IBM 1130 FORTRAN Primer*. Scranton, Pa.: International Textbook Company, 1969.

Pennington, Ralph H., *Introductory Computer Methods and Numerical Analysis*. New York: Macmillan, 1965.

Veldman, Donald J., *FORTRAN Programming for the Behavioral Sciences*. New York: Holt, Rinehart & Winston, 1967.

Index

A1 format   115
A2 format   116
A3 format   117
A2 format vs. A3 format   119
A conversion: *see* Alphanumeric conversion
Accuracy of calculations   137
Adjustable dimensions: *see* Variable dimensioning
Alphanumeric conversion   9
  subject identification fields   30
Alternate stacker   113
Analytic rotations   60
Argument lists   66
Arrays
  compressed   72
  disk resident   93
  names   131
  nonsymmetric   72
  printing   7, 11, 71, 74, 78
  symmetric   75
  temporary storage on disk   89
  variable dimensioning   78
Associated variable   82, 84

Biomedical computer programs   38
Blank card: *see* Termination of execution with blank card
Blocking data: *see* Disk
Borko, H.   vii

CALL LINK   62
  multiple entry points   70
  passing temporary files between mainlines   86, 97
  permanent disk files   106
  program C01   135
  *STORE control record   98
Character strings   108
Characteristics of library programs   3
Characters
  commercial subroutine package   107
  double precision   31
  integer array elements   21
  packed   116
Comment cards   130
COMMON
  associated variable   84
  dynamic storage area   63

renaming an area 63
subroutines 25
use in mainline-to-mainline chaining 62
versus argument lists 66
Compressed storage: *see* Arrays
Computational notes in documentation 48, 56
Computer language, Choice of 4
Consultation with users 36
Control cards 3
  descriptions 40
  program 125
  schemes 5, 9, 20, 21, 23, 25, 26, 125
  standardization 124
  system 9, 124
Conversion 6
  A1 format to A2 format 116
  A1 format to A3 format 118
  A2 format to A1 format 117
  A3 format to A1 format 118
  alphanumeric to numeric 9, 27
  numeric to alphanumeric 111
  programs from one computer to another 2
Cooley, W. W. vii
Core map 103
Core memory 60, 65
Core-resident software 67
Covariance 137
CSP: *see* IBM Commercial Subroutine Package

Data format 4
Data processing consultant 2, 36
DCI: *see* Disk core image format
Deck identification code 132
DEFINE FILE Statement 82, 85, 97
Deviation scores 136, 138, 152
DFILE 105
Diagnostics 4
Digital Equipment Corporation 67

PDP-11 16, 59
Disk
  blocking data 85
  extension of core memory 93
  files, permanent 49, 103
  files, permanent, and library programs 105
  files, temporary 81, 97
  input from disk 49
  I/O lists that exceed 320 words per subject 87
  organization 82
  READ statement 84
  record length 85
  sector 82, 104
  storing data as one record per subject 84
  storage of matrices 93, 96
  storage of programs 98
  user area 98
  working storage 98
  WRITE statement 83
Disk core storage format 99
Disk storage format 98
Disk utility program 144
Dixon, W. J. 38
Documentation 4
  descriptions of programs 128
  identification of programs 12
  sample problem and output 49, 56
Double precision 31
DSF: *see* Disk storage format
/ / DUP: *see* Disk utility program

Echo checks 12, 30
Embedded blanks 10, 17
Entry points: *see* CALL LINK
Error checks 14
Error messages 4, 9, 12, 14, 124
Executable files 124

Files: *see* Disk

*FILES 105
FIND statement 91
  use with program segmentation 92
Floating point 67
Flow charts
  program C01 140
  program C01A 145
  program C01B 148
  program C01C 150
  program D03 157
  program D03A 161
  incorporating a user-supplied subpro-
    gram 100
  organization of program C01 136
  organization of program D03 155
  phases of processing 69
  principal components analysis 61
  separating forms of I/O 115
  subroutine START 9
// FOR 139, 189
Format cards 43, 45, 47
Format specifications stored in an
  array: see Variable format
FORTRAN 1, 4, 67
Frequency tabulation 155

Generalizing a program 5
George Peabody College Computer
  Center 186
Gruenberger, F. 67
Guilford, J. P. 76

Headings, Generation of integer 7
Hewlett-Packard Company 31, 67
Hewlett-Packard 2100 series computers
  59
Hogge, J. H. 38
Horst, P. viii

IBM 360 31, 82
IBM 1130 47, 59
  disk 81

executable files 124
FORTRAN 16, 21, 63, 132
  integer arithmetic 67
  mainline-to-mainline chaining 64, 71
  operating system 62, 65
IBM 1442 card read/punch 47, 102, 108,
  110, 112
IBM 2315 disk cartridge 82
IBM 2501 card reader 109
IBM Commercial Subroutine Package
  107, 195
  input/output 107
  restrictions concerning input/
    output 113
IBM Scientific Subroutine Package 195
IBM Statistical System 196
Identification sequence: see Sequence
  numbering
Incomplete data: see Missing data
Input
  control cards 9
  device numbers 6
  standardization 124
  see also Variable format
Integer arithmetic 67
Intercorrelation matrix 60
Interfacing 105
Intermediate output 60
*IOCS 115, 140, 147, 191
Item location cards 30

Jargon 4, 36
// JOB 189

Language features 132
Larson, C. 67
Library program sources viii, 3
*LIST SOURCE PROGRAM 139,
  190
*LOCAL 65, 92, 99, 136, 144
Lohnes, P. R. viii
Louden, R. K. 113

Magnetic tape 60
Main program: *see* Mainline
Mainline 60, 68
Mainline-to-mainline chaining 64, 70, 135
  see also CALL LINK
Mass storage 26, 60, 81
Matrices: *see* Arrays
Mean 6, 136
Minimization of deck size 25
Missing data 22
Missing data signal card 23
Multiple entry points: *see* CALL LINK
Multiple format cards 14
  simulated variable format 17
Multiple measurements 126

Need for library programs 1

Object-time format specification: *see* Variable format
*ONE WORD INTEGERS 113, 139, 189
Operating systems 59
Output
  device numbers 6
  standardization 127
Overlays of subprograms: *see* *LOCAL

Parameter cards 9, 10, 47, 55, 125
Peabody Statistical Library 196
Peabody Statistical Library User's Manual 38
Peabody Time Accounting System 193
Pearson product-moment correlation 135, 137
Pennington, R. H. viii
Perry, G. L. 133
Phases of processing 67
Picklesimer, J. H. 38
Pollack, S. V. viii
Precompiled library programs 16

Principal components analysis 60
Problem parameters 4
Problem segmentation: *see* Segmentation of programs
Program libraries 1
Program output 127
Program segmentation: *see* Segmentation of programs
Programs and subprograms
  program C01 45, 135, 139
  program C01A 144
  program C01B 147
  program C01C 149
  program D03 53, 155, 156
  program D03A 160
  subroutine A1A3 118
  subroutine A3A1 118
  subroutine DCODE 27, 168
  subroutine DCPT 94, 169
  subroutine DCPTN 95, 169
  subroutine DDPCH 153, 170
  subroutine DPNCH 171
  subroutine DPRT 78, 172
  subroutine DPRTN 78, 91, 172
  subroutine ECODE 111, 173
  subroutine FMAT 18, 139, 174
  subroutine MISS 22, 176
  subroutine MPRT 11, 74, 97, 177
  subroutine MPRTN 91, 178
  subroutine PACK 116
  subroutine PUNCH 110
  subroutine PUNCH vs. WRITE 112
  subroutine R2501 109
  subroutine READ 108
  subroutine ROUND 152, 178
  subroutine STACK 113
  subroutine START 9, 139, 178
  subroutine TAKE 19, 179
  subroutine TAKEC 110, 147, 180
  subroutine TAKEI 32, 181
  subroutine UNPAC 117
  subroutine XTRAN 101

Punched output 47

Ralston, A. viii
READ statements 67
Real arithmetic: *see* Floating point
Rounding 152
Run-time format: *see* Variable format
Running summation 7

Sample problem: *see* Documentation
Scratch storage 89
Scratch tape 22
Sector: *see* Disk
Segmentation of programs 60
Segmentation of an array in printing:
   see Arrays
Separation of phases of processing 70
Sequence numbering 26, 132
Serial order of input 21
Sigma 137
Simulated variable dimensioning: *see*
   Variable dimensioning, Simulated
Simulated variable format: *see* Variable
   format, Simulated
SOCALs 103
Software implications of FORTRAN
   statements 67
Sommerfeld, J. T. 133
Source deck standardization 130
Source statements 132
Specification statements 83
Stacking data sets 16, 26
Statement numbers 132
Statistical consultation 36
Sterling, T. D. viii
Storage of variable names 91
*STORE 98, 193
*STORE vs. *STORECI 102
*STORECI 99, 125, 144
*STOREDATA 104

Subject identification fields 30, 126
   simulated variable format 32
Subprograms and subroutines: *see* Pro-
   grams and subprograms
Subroutine names 26
Subscripting 74, 76, 80
System mass storage 16

Tape operating systems 64
Termination of execution with blank
   card 10
Title card 9
Title of program 12
Trailing blanks 17

Use of blank control card: *see* Termina-
   tion of execution with blank card
User area: *see* Disk
User errors 4, 8, 14
User-supplied subprograms 100
User training 35
Users, Assumptions about 35
   attitudinal predisposition 36
   prior experience 35
   statistical competency 36

Vanderbilt University Computer Center
   123
Variable dimensioning 78
Variable dimensioning, Simulated 79
Variable format 11, 15, 16, 73, 75
Variable format, Simulated 16, 73,
   76, 110
Variable location cards 26, 53
   subject identification fields 31
Variable names 21, 78, 131
Variable names cards 125
Variable numbers 7
Variance 137
Veldman, D. J. viii

Version date   12   WRITE statements   67

Wilf, H. S.   viii   // XEQ   191
Working storage: *see* Disk